water rites

reimagining water in the west

Calgary Institute for the Humanities Series
Co-published with the Calgary Institute for the Humanities
ISSN 2560-6883 (Print) ISSN 2560-6891 (Online)

The humanities help us to understand who we are and where we came from; they help us to understand and respectfully engage with those who are different from us; and they encourage a curiosity and imagination that allows us to bring older ideas to the new worlds in which we find ourselves. Books in this series embody this spirit of inquiry.

No. 1 **Calgary: City of Animals**, edited by Jim Ellis
No. 2 **Water Rites: Reimagining Water in the West**, edited by Jim Ellis

edited by jim ellis

water rites

reimagining water in the west

a special publication of the university of calgary press
in co-operation with the calgary institute for the humanities

CALGARY INSTITUTE FOR THE HUMANITIES SERIES
ISSN 2560-6883 (Print) ISSN 2560-6891 (Online)

University of Calgary Press
2500 University Drive NW
Calgary, Alberta
Canada T2N 1N4

press.ucalgary.ca

Library and Archives Canada Cataloguing in Publication

Water rites : reimagining water in the West / edited by Jim Ellis.

(Calgary Institute for the Humanities series, ISSN 2560-6883, ISSN 2560-6891)
"A special publication of the University of Calgary Press in co-operation with the Calgary Institute for the Humanities."
"The following collection of essays, commentary, and artworks about both water rights and water rites has its origin in the Calgary Institute for the Humanities' Annual Community Seminar of 2017."—Page xvii.
Includes bibliographical references.
Issued in print and electronic formats.
ISBN 978-1-55238-997-3 (softcover).—ISBN 978-1-55238-998-0 (Open Access PDF).—
ISBN 978-1-55238-999-7 (PDF).—ISBN 978-1-55238-934-8 (EPUB).—
ISBN 978-1-55238-935-5 (Kindle)

1. Water-supply—Social aspects--Alberta. 2. Right to water—Alberta.
3. Water conservation—Alberta. 4. Alberta—Environmental conditions.
I. Ellis, Jim, 1964-, editor II. Calgary Institute for the Humanities, issuing body III. Series: Calgary Institute for the Humanities series

TD227.A4W38 2018 333.91'16097123 C2018-901934-4
C2018-901935-2

The University of Calgary Press acknowledges the support of the Government of Alberta through the Alberta Media Fund for our publications. We acknowledge the financial support of the Government of Canada. We acknowledge the financial support of the Canada Council for the Arts for our publishing program.

Canada Council for the Arts Conseil des Arts du Canada

Printed and bound in Canada by Marquis
This book is printed on Hannoart Plus 100 paper

Copyediting by Kathryn Simpson
Front & Back flap image: *Man and Woman on Sandbar in Elbow River*, Calgary, Alberta.
Glenbow Archives, PA- 3689-71, Calgary, AB.
Cover art: Leslie Sweder, *Last Light, SW Corner of Confluence, July 18, 2017*

Cover design, page design, and typesetting by glenn mielke

The Meeting of the Bow and High Rivers, N.W.T. by Thomas Bland Strange, c. 1882, hand-coloured engraving on paper, Glenbow Archives, 61.81.17, Calgary, AB.

contents

acknowledgements

This book originated in the Calgary Institute for the Humanities' Annual Community Seminar for 2017. The idea for the seminar came from the CIH's Advisory Board, which included Bill Dickson, Jackie Flanagan, Murray Laverty, Donna Livingstone, Judy MacLachlan, Bill Ptacek, Valerie Seaman, Nancy Tousley, and Rod Wade. Many thanks are due for their guidance and support of the project from beginning to end; I am particularly grateful to Nancy Tousley for her advice on art and artists.

At UCalgary Press, Brian Scrivener, Helen Hajnoczky, Alison Cobra, and Melina Cusano have all been very supportive of the project and our vision for the book; thanks to Glenn Mielke for the beautiful design.

At the CIH, Sharla Mann and Moe Esfahlani provided research assistance and organizational support for the community seminar. Sharla Mann, outgoing coordinator of the CIH, did picture research and provided administrative support for the early stages of the book with efficiency and good humor; Sean Lindsay, the incoming coordinator, helped in the final stages of book production. I am grateful to the three original panelists of the seminar, as well as to all of the other contributors to the book, for being co-operative and patient with our many requests, as the outlines of the book came into shape. A heartfelt thank you to all of the writers, artists, and organizations who allowed their work to be included.

Next page: **Canoes in a Fog, Lake Superior, 1869** by Frances Ann Hopkins, 1869, oil on canvas, Glenbow Archives, 55.8.1, Calgary, AB.

introduction

rethinking our relations to water

jim ellis
director, calgary institute for the humanities

The city of Calgary, like many sites of human habitation, is located on a body of water. Or in our case, two: for centuries, there have been settlements and human activity at the confluence of the Bow and the Elbow rivers, a place that the Blackfoot peoples called *Mohkínstsis*. Humans settle by water for numerous reasons: most immediately, because a source of clean drinking water is necessary for survival, but also because water allows for fishing, agriculture, industry, and frequently transportation. Water is a potent symbol in so many cultures and religions precisely because it is so foundational to life. Water connects everything on the blue planet: the human, the animal, and the material world. And in the face of global climate change, water is an increasingly scarce and precious resource.

Here in Western Canada, we live in a vast watershed that extends from the Rocky Mountains, passes over the prairies, and flows into James Bay, crossing all kinds of human-made boundaries. As David Laidlaw points out in his essay in this volume, water issues are particularly acute in the West, which is experiencing more regular cycles of droughts and floods; the forest fire season gets longer every year; the glaciers continue to recede at an alarming rate; and logging on the mountain slopes, along with the damage caused by recreational activities, decreases the ability of the watershed to retain water, resulting in greater flooding downstream. All of these water issues are connected, and point to our own connectedness: the flow of water connects human activity all along its course, affecting all life forms downstream — whether they be human, animal, or plant. Water visibly reminds us of our connections, as well as our responsibility, to those that share the same watershed.

On July 28, 2010, the United Nations "explicitly recognized the human right to water and sanitation and acknowledged that clean drinking water and sanitation are essential to the realisation of all human rights" (see Appendix A). There have been notable failures in our country to provide access to this resource, particularly in our northern Indigenous communities.[1] On March 22, 2016 — World Water Day — the Canadian government committed to addressing the infrastructure problems that have plagued these communities; they have pledged, through Indigenous Services Canada, to end all water advisories for systems it finances by March 2021.[2] Addressing physical infrastructure is a key issue, but these past failures are linked to larger systemic problems identified in the final report of Canada's Truth and Reconciliation Commission (TRC).[3] Again, we can see some positive movement: after years of delay, the Canadian government has finally removed its objector status and adopted the UN Declaration on the Rights of Indigenous Peoples. In Appendix B, we reproduce some of the relevant articles of the declaration concerning water, the environment, land use, and development; among other things, the declaration upholds the rights of Indigenous communities to grant or withhold consent to development on their territories. The issue of free, prior, and informed consent is a thorny one for communities facing chronic economic hardships, as Michelle Daigle observes in her essay below. Resource extraction and industrial development can both contaminate and exhaust water supplies, and frequently our thinking is focused too narrowly on economic benefit, and is too shortsighted about ecosystem timelines.

Both the TRC on the one hand, and the environmental movement on the other, challenge us to consider not just the way we have managed and mismanaged our water, but how we have thought about it, and how we might balance competing claims on water in the future. How can we rethink our relation to water? While stressing the human right to water, do we need to think at the same time about our responsibility to water? Do non-human entities such as fish, plants, and water have rights? What might different spiritual or cultural traditions tell us about our duty to water? What might our duty to water tell us about our connection to each other?

Within Indigenous communities, bodies of water are often seen as beings with their own inherent rights and status. The writer and Sto:lo elder Lee Maracle writes: "We do not own the water, the water owns itself."[4] As Nancy Tousley mentions in her discussion of the artist Tanya Harnett, this way of thinking was tangibly recognized in a groundbreaking legal decision when, in March 2017, the New Zealand legislature granted the Whanganui River rights as a living entity. This decision came in response to a long campaign waged by the Maori community, who see the river — which they call Te Awa Tupua

— as a spiritual being. Just days after that decision, a court in India recognized the Ganges and the Yamuna River as living entities with the same rights as persons and in the following weeks, similar status was granted to Himalayan glaciers. What does it mean to grant a body of water rights? Or, to shift the question slightly: what does it mean to recognize the rights of water?

While we have not yet had a similar legal finding in Canada, on December 1, 2017 the Supreme Court did rule in favour of First Nations and environmental groups who were fighting the Yukon government's plan to allow more mining in the Peel watershed. The court's decision highlighted, among other things, the failure of the government to respect treaty obligations with the Tr'ondëk Hwëch'in, Na-Cho Nyak Dun, and the Vuntut Gwichen First Nations. This decision was counterbalanced not long afterwards, when on December 11, 2017, the new government of British Columbia decided to go ahead with the massive hydroelectric project known as the Site C dam, in spite of much resistance from Indigenous and environmental groups. Helen Knott, who writes below about her experience as an Indigenous activist, participated in protests against this project, which will flood the territory of the West Moberly and Prophet River First Nations in the Treaty 8 region. Clearly, Canada has some distance to travel in readjusting our thinking about water rights; addressing these very urgent water issues requires us to go beyond scientific data and economic projections to consider the multitude of voices and perspectives that reflect water's omnipresence in our lives, our thinking, and our imagination.

The following collection of essays, commentary, and artworks about both water rights and water rites has its origin in the Calgary Institute for the Humanities' Annual Community Seminar of 2017. The Calgary Institute for the Humanities is Canada's oldest humanities institute, founded at the University of Calgary in 1976 to support and promote the values of humanities research. Each year the advisory council of the CIH chooses a topic for discussion that is both timely and important for the larger communities that we serve; this was the second seminar in a planned trilogy on issues drawn from the environmental humanities. The seminar, "Water in the West: Rights of Water / Rights to Water," had three guest speakers — Michelle Daigle, David Laidlaw, and Adrian Parr – who, along with moderator Tasha Hubbard, explored these issues with a group of seventy members of the Calgary community.

The assembled participants included representation from, among others, Calgary's Aboriginal Urban Affairs Committee, City of Calgary Parks Department, the Watershed+ resident artist program of the city's water department, the Tsuu T'ina First Nation

Environmental research department, Bow Basin Watershed Management Plan, Alberta Ecotrust, Elbow River Watershed, Yellowstone to Yukon Conservation Initiative, Alberta First Nations Information Governance Centre, and the Alberta Wilderness Association. The day began with a blessing from Elder Wallace Alexson, followed by opening statements from our three guests that we have included in revised form in this book. Over lunch, different table groups discussed questions that were posed by the speakers, and these groups prepared responses that formed the basis of the afternoon's discussions. In addition to the talks by our three speakers, many of the other contributions to this book came from seminar participants.

The day also featured a screening of the film, *Gloire à l'eau* (1935/1950) by the Quebec amateur filmmaker Albert Tessier. This film begins with images of baptism, a water rite that marks the entry into a spiritual community. As Charles Tepperman notes in his discussion of the film, Tessier was both a priest and an early environmentalist, and his film explores some of the same aspects of water that this volume will explore: the sacred and spiritual significance of water; its necessity as a source of life and a provider of food; its use as a transportation route for people and goods; and its restorative role as a site of recreation and source of aesthetic inspiration and contemplation.

This collection echoes Tessier's film in its attempt to offer a similarly broad spectrum of perspectives on our relationship to water: from discussions of the most effective form of rights-based arguments; to the history of how water rights in Alberta, from contact onwards, were negotiated; to thinking about contemporary examples of how resource extraction or power generation impact Indigenous communities. Contributors discuss the activities of activists and environmentalists in protecting crucial watersheds and protesting harmful developments. And we include a series of contributions by and about artists, both historical and contemporary, who have engaged with water in different ways. Underlying much of this is a question of values, and the way that different cultures have entertained radically different approaches to certain concepts — such as land ownership — or entities — such as water.

Adrian Parr, a UNESCO Co-Chair of Water Access and Sustainability at the University of Cincinnati, questions whether a rights-based approach to water access is sufficient for political change. Looking closely at the water crisis in Flint, Michigan and moving on to consider the protests at Standing Rock, Parr draws attention to how access to water is often bound up with larger issues of class, race, and gender as well as income disparity. Ultimately, she calls for more attention to a common right to water and coalitional approaches to political action. David Laidlaw, a research associate at the Canadian Institute of Resources Law at the University of Calgary, offers a detailed legal history of Indigenous water rights in Alberta, particularly as they were established in the numbered treaties, and how these have been affected by subsequent legislation. He outlines the history of how water has been allocated by the province of Alberta, more recent attempts to deal with water scarcity, and the increasing challenges that the province will face in coming years due to climate change.

Like water, some of these essays move across traditional boundaries. The essays by Michelle Daigle and Helen Knott combine academic approaches with family history and personal reflections. Daigle, an assistant professor in the Department of Geography at the University of British Columbia, writes of her grandparents' experience in Treaty 9 territory (in the area known as the Ring of Fire), and the effects of resource extraction on First Nations communities. Daigle details the challenges that First Nations communities often confront when offered the difficult choice between much-needed economic development and possible environmental degradation and loss. Her essay addresses the need for a wider perspective that acknowledges how water crosses boundary lines and the kinds of relationships that water creates. Knott, a First Nations writer and activist, talks about her experience in organizing protests against the construction of the Site C hydroelectric dam in Northern British Columbia. Knott's essay crosses the boundary between academic and activist writing. It also provides a bridge to some of the contributions to this volume by community-based groups such as Yellowstone to Yukon Conservation Initiative, Alberta Ecotrust, and the Elbow River Watershed Protection group, who are involved more directly in the attempt to preserve and protect water sources and watersheds, and to educate those who swim (literally and metaphorically) in these water systems.

In attempting to shift our thinking about water, it is important to address not just how it figures in our daily lives, but how it lives in our imagination. To that end, this book contains art criticism by Nancy Tousley and Ciara McKeown, as well as artists' statements and portfolios of work by painter Leslie Sweder and photographer Warren Cariou. Tousley discusses the work of Tanya Harnett, whose "Scarred/Sacred Water" viscerally exposes the wounded waterscapes in northern First Nations communities. McKeown's essay on Watershed+ details an innovative approach to public art that embeds artists in Calgary's water department. Watershed+ is an ongoing program that fosters collaborations between artists, planners, and engineers, producing innovative art that deepens our understanding of our own immersion in the watershed.

In an ongoing series that spans several years of practice, Sweder paints the surface of the Bow River *en plein air*, near where she lives at the confluence of the Bow and the Elbow rivers. The project, *Notes on Current*, is simultaneously an artistic experiment and an act of loving attention to her surroundings. A parallel project not included here, *The Things We've Left Behind*, photographs the flotsam and jetsam that Sweder observes on the shore or in the water — such as an abandoned bicycle luxuriantly covered with algae. Cariou's photographic practice is more directly political, using bitumen gathered on the banks of the Athabasca to develop "petrographs" of the Alberta tar sands. The process of developing these photographs using bitumen is, says Cariou, highly toxic; this thoughtful use of naturally occurring bitumen to expose the toxic realities of bitumen extraction is, like Harnett's ritual practice, a performance — or indeed, a rite — that helps to reorient our thinking about our responsibility to water sources.

The artworks that we reproduce, both historical and contemporary, are important for showing how water has figured in our imagination, and how it has helped to shape the cultural imaginary, particularly in the West. We see water as a primary conduit of colonization in Frances Anne Hopkins' "Canoes in Fog, Lake Superior" (1869). Similarly, in the engravings executed by Thomas Strange for the nineteenth-century magazine *Canadian Illustrated News*, we see the Bow River centrally featured as a locus of activity and settlement. In his "Blackfoot Crossing" (1882), Strange puts a solitary First Nations figure in the foreground, and in the background we see both Indigenous encampments and the first signs of European settlement. In the painting "Morning, Lake Louise" (1889) by Frederic Marlett Bell-Smith, water is a spectacle in itself, an iconic image that can be used to attract others to the region, and a key part of how we imagine ourselves. Shelley Ouellet responds to this use of water as tourist spectacle in her monumental installation *Wish You Were Here* (2001). In an essay accompanying this exhibition, Amy Gogarty notes that Bell-Smith painted Lake Louise less than ten years after the first European sighting, and that almost immediately, "the Canadian Pacific Railway commandeered its spectacular beauty for commercial purposes, facilitating a steady stream of tourists well before the end of the [nineteenth] century."[5] For her installation, Ouellet fabricated huge glittering curtains (96 x 180"), using black, white, and clear plastic beads to reproduce three iconic Canadian waterscape images. These nineteenth-century paintings of Lake Louise, Niagara Falls, and the Saguenay River helped to frame Canada's self-identity as a land of spectacular, pristine, inexhaustible natural resources. The larger tradition of landscape painting from which they come invokes a romantic approach to nature, positioning the heroic individual against a hostile or threatening Nature that must be tamed, subdued, or exploited. In response Ouellet employs a collaborative, community-based practice that challenges Western ideas of the solitary and heroic individual, whether artist or explorer.

Exploring the way that artists have responded to water and water issues is a crucial part of the response to the human right to water and the challenges of global climate change. Science can offer proof of climate change, the humanities can explore and expose its human dimensions, and art persuades us on a different — and arguably more fundamental — level, intervening in the imagination. All of these approaches are necessary and complementary. The range of voices and images in this collection together aim at shifting our understanding not just of the role water plays in our lives, and the consequences of our misuse of it, but more fundamentally of the way that water connects us all, on every level. We are all bodies of water.

notes

1. A Government of Canada Advisory bulletin notes that as of November 2017 there were ninety-five long-term advisories and forty-one short-term advisories in place in First Nations communities south of the 60th parallel. This did not include advisories in British Columbia, of which there were twenty-two. https://www.canada.ca/en/health-canada/topics/health-environment/water-quality-health/drinking-water/advisories-first-nations-south-60.html.

2. Matthew McClearn, "Is a Lack of Training Hindering Progress on Water Advisories?," *Globe & Mail*, January 30, 2018, A10–11.

3. *Make it Safe: Canada's Obligation to End the First Nations Water Crisis*. Human Rights Watch, June 2016, 4. This report offers a comprehensive introduction to the water access problems in Canada's Indigenous communities, along with a series of recommendations. https://www.hrw.org/report/2016/06/07/make-it-safe/canadas-obligation-end-first-nations-water-crisis.

4. Lee Maracle, "Water," in *Downstream: Reimagining Water*, eds. Dorothy Christian and Rita Wong (Waterloo: Wilfred Laurier University Press, 2017), 37.

5. Amy Gogarty, Essay accompanying the exhibition, Shelley Ouellet: *Wish You Were Here*, Nickle Arts Museum, February 8–March 30, 2002.

Arthur Moore (2008) by Elizabeth Moore.

embodying kinship responsibilities in & through nipi (water)

michelle daigle

Mushkegowuk *(Cree), member of Constance Lake First Nation*
assistant professor, department of geography,
university of british columbia

knowing nipi[1] through embodied and storied practice

Angela and Bert Moore, my *kokom*[2] and *moshoom*,[3] were born in the 1920s on the muskeg lands and waters of our nation: the *Mushkegowuk* Nation, now known as Treaty 9 territory. At the beginning of summer each year they would leave their respective communities, where they resided during the winter months, and travel with their families along *Kishiichiwan*, otherwise known as the Albany river.[4] They would drink water directly from the river, fish for pickerel and pike, snare rabbits, pick berries, and towards the beginning of fall, hunt for ducks and geese. Throughout the summer, they would visit many place-names on *Kishiichiwan*, and they would gather with other *Mushkegowuk*, *Anishinaabe,* and Oji-Cree families at "the forks," where *Kishiichiwan* meets the *Kenogami* River. There they would visit with their relatives, get updated on local news and gossip, discuss pressing political concerns, and renew diplomatic relationships.

It was through this time on *Kishiichiwan*, through embodied experiences, that every summer my grandparents learned their responsibilities to *sibi*,[5] and more generally to *nipi*. Simultaneously, they learned their responsibilities to *Mushkegowuk* kin who depend on *Kishiichiwan*. These are kin that they came to know while being out on *Kishiichiwan*, including animal and plant kin. It is how they came to understand that what they learn from these kin, as well as their responsibilities to these kin, make up *Mushkegowuk* forms of governance: governance practices that extend much further back than colonial settlement, and which continue to be renewed into the present and future.

Although it became increasingly difficult, my *kokom* and *moshoom* maintained their relationship with *Kishiichiwan* despite the increasing impacts of colonial capitalism in our home territory. They continued to renew their relationship despite being forced to attend residential school,[6] and despite relocating 300 kilometres south of the James Bay area — where they grew up — to a newly vested reserve in Treaty 9, Constance Lake First Nation (CLFN). They left the James Bay area in search of new employment; many Indigenous families were experiencing the economic impacts of the end of the fur trade era. New jobs were becoming available a few hundred kilometres south in the booming lumber industry, and in the infrastructural development of the Trans-Canada Highway and the Canadian National Railway. My *moshoom* worked in the pulp and paper mill and, later on, the lumber mill that was established in CLFN, which supplemented his income from fur trapping that he continued to do during the winter. Meanwhile, my *kokom* increasingly became tied to the domestic sphere as she took on sole responsibility for their children: my *moshoom* spent extensive time on their trapline during the winter, and worked long hours in the lumber mill during the summer months. In addition to this reproductive labour, my *kokom* engaged in seasonal employment, such as tree planting, and eventually found a more permanent position as a Cree language teacher in a public school in a nearby town.

In some ways, their everyday lives were becoming much different from those of their parents and grandparents; they became increasingly dependent on the growing settler Canadian economy. Yet they returned to *Kishiichiwan* whenever they could. In some instances, they travelled down *Kishiichiwan* to visit their children at St. Anne's residential school (they could not afford plane tickets). In other instances, during the summer break, they would bring their children on *Kishiichiwan* as their parents had once brought them. Extended travels, however, remained difficult due to their employment obligations and the geographic distance of *Kishiichiwan* from CLFN. They started building new relationships in the place where they now resided, spending a great deal of time on the *Kabinakagami* river, a river that is part of the Albany river watershed, and which flows through CLFN. Eventually they built a cabin on the *Kabinakagami* River. This is the river that I and their grandchildren came to know through fishing, hunting, swimming, berry picking, tending to the garden my grandparents planted at the cabin, and drinking *wabaanomakun* (tea pulozz) long into the evening hours.[7]

My *moshoom* passed on into the spirit world in 2008, at the age of eighty-two. Many of the stories shared about him during the grieving time took place on the *Kishiichiwan* and *Kabinakagami* rivers. Family member shared stories they had been told about people's first encounters with my *moshoom* on *Kishiichiwan*, as he travelled down the river with his parents as a newborn. Meanwhile, my *kokom* told me about the times she and my *moshoom* paddled from Fort Albany, down to the area where CLFN was eventually established, to visit family and friends who had already relocated to one of the many settlements that existed in the area before Indigenous peoples were forced to live on reserves. As a young adult, these stories reminded me that my family and nation's stories flow through these rivers. My *moshoom* was teaching me this, even as his physical presence was no longer with us.

A few years passed before I started asking my *kokom* for more stories about the *Kishiichiwan* and *Kabinakagami* rivers. When I did, her eyes lit up with love and excitement. Whereas she once taught me while being out on *aski*[8] — at the age of ninety-one — she now teaches me through her stories. I travel home from *Musqueam*, *Squamish*, and *Tsleil-Waututh* territories, otherwise known as Vancouver, British Columbia, where I now work and live as a Cree visitor, and I sit with her. I sit and I listen.

Her stories have taught me that *nipi* is our kin, a relative and legal actor with whom we must renew our relationship, just as we must renew relations with our human kin. Without a direct and intimate relationship with *nipi*, how can we continue to be in good relation? "You need to go there," my *kokom* tells me.[9] In this way, her stories complicate the notion of human rights to water that has overwhelmingly framed water security and governance policy-making within national and international institutional forums. Indeed, Indigenous communities across the globe, including those in Treaty 9, continue to fight for water security and governance by mobilizing a human rights discourse. At the community level, however, many Elders and Knowledge Holders who are the legal caretakers of water according to Indigenous laws and political orders, continue to centre the *responsibilities* and *accountabilities* we have to our kin, *nipi*.[10] Further, her stories stress the accountabilities we have to our animal, plant, and human relatives that depend on *nipi*, and that we are temporally and spatially connected to through our waterways. Specifically, they teach us that the everyday and intimate relationships that we build in and through water, through embodied and storied practice, ripple out through time and space. Our relational accountabilities encompass this vast kinship network.

nipi is sick

My *kokom*'s stories also teach me about how our kinship relations have been impacted through capitalist dispossession and violence. She, like many other Indigenous peoples, is witnessing, embodying, and testifying to the ways our sacred waters are increasingly becoming entangled in extractive developments. It is common to hear from Elders and community members that *nipi*, as well as pike, pickerel, deer, and moose, are sick because of these developments, just as we would describe how human relatives experience sickness. These are people who have intimate relationships with *aski*, through land and water-based practices such as hunting, fishing, and trapping. Throughout the years, they have observed changes in animals such as deer — notably changes in their livers — as well as changes in local vegetation. As my *kokom* says, such shifts are all connected to the contamination and pollution of the Mushkegowuk waters that these animals and plants depend on.

Currently, the Albany and Attawapiskat watersheds are entangled in mining developments. The De Beers Victor mine has been operating outside of Attawapiskat First Nation since 2008. Furthermore, a proposed mining development dubbed the "Ring of Fire" is reported to be the largest chromite deposit in North America, with an estimated value of 30 billion dollars.[11] If approved, the project will span 5,000 square kilometres. Additionally, a graphite deposit, now known as the Albany deposit, or the "Arc of Fire," has been identified on the lands of CLFN. By 2015, nearly two dozen companies held claims in Mushkegowuk territory, and 278 million dollars was spent in exploratory drilling alone.

Due to mining activity at the De Beers Victor mine and initial exploratory drilling for the "Ring of Fire," mercury and methylmercury are on the rise in the Albany and Attawapiskat rivers. This is disproportionately impacting caretakers of these rivers, including pike and pickerel, as well as Indigenous women and children. Indigenous women in the area are experiencing higher cases of infertility, miscarriages, and birth deformities, as *nipi*, herself experiencing violence, is unable to fulfill her responsibilities to give and support life. Hence, the first place where Indigenous women, children, and water come into relationship — the womb — has become a particularly precarious place, even as it still, always, embodies love and strength.

Mining is simply the most recent instance of dispossession and violence in a much more extensive history of resource extraction that has contaminated and poisoned *Mushkegowuk* waters. Some scientists have connected the rising levels of methylmercury, in the Albany and Attawapiskat rivers, to mercury that was originally generated through the pulp and paper and lumber industries. Specifically, mercury deposits generated through the lumber industry are released through new mining activity that generates methylmercury, a neurotoxin that threatens the health of human and aquatic life. De Beers has strategically deployed the lumber industry's connection to mercury in Treaty 9 to evade legal and economic accountability to Indigenous communities. The company has continued to use this corrupt rationale to deny all accountability to the ecological, infrastructural, and health hazards that the Victor Mine has caused in and around Attawapiskat First Nation. This continues to happen despite the people of Treaty 9 demanding accountability from De Beers and the Canadian government.[12]

At the same time, the impact of extractive industries is directly tied to the contamination of clean and safe drinking water in Treaty 9. The Canadian government's longstanding colonial and racist refusal to economically support the infrastructural development of water treatment facilities within Indigenous communities across the country further exacerbates this negative impact.[13] Consequently, many Indigenous peoples have and continue to outsource, and even purchase, their drinking water. During my childhood, for example, my grandparents would come to our home every Saturday, which was located in a town approximately thirty kilometres away from CLFN. They would fill up water jugs, and this would be their drinking and cooking water for the week.

Many Indigenous families across Canada outsourced their water for decades throughout the twentieth and twenty-first centuries, and many continue to do so today. In June of 2017, 132 water advisories were in effect across Indigenous communities in the country (Health Canada 2017).[14] This is a conservative estimate given that reports do not include advisories in Indigenous communities in British Columbia, and within the Saskatoon Tribal Council. Furthermore, the Canadian government limits its reporting to Indigenous communities south of the 60th parallel, thus excluding communities in northern parts of Nunavut, the Northwest Territories, and the Yukon Territory.

Water deficits in Treaty 9 communities, and accompanying colonial and racialized narratives of Indigenous communities as poverty-stricken sites of individual failings and physical and social decay, have become central to the ways the Canadian government, industry, and mainstream media legitimate proposed mining developments. Mining, they argue, will serve as a solution to poverty and the lack of infrastructure within Indigenous communities — including the lack of water treatment facilities. Yet these technocratic solutions rarely generate jobs and new flows of capital in Indigenous communities. Instead, mining companies increasingly use Impact Benefit Agreements (IBAs) to limit their accountability to the capital and employment they promise to Indigenous communities. Moreover, IBAs are used to discipline and sanction Indigenous resistance when such agreements are not met. Hence, many Indigenous peoples of Treaty 9 understand mining as a continuation of structural colonial power relations and Indigenous land dispossession.

Pushbacks against colonial technocratic "solutions" are reinforced by community members' understandings of interconnected and relational ecologies. For example, in 2016 CLFN opened up its first water treatment facility. Many Elders in the community, however, continue to buy filtered bottled water from a nearby town. Initial conversations suggest that Elders continue to outsource their drinking water due to a fundamental mistrust in state-sponsored strategies that is, in turn, rooted in years of witnessing and experiencing colonial dispossession. Furthermore, many community members stress how *Mushkegowuk* lands and waters continue to be contaminated and polluted through unsustainable resource extraction, thus contextualizing clean and safe drinking water within larger regional environmental issues (not to mention legal issues, since these extractive developments continue without Indigenous peoples' full and informed consent).

This is not to state that access to clean and safe drinking water is not crucial, especially given that Indigenous communities continue to face legal, financial, and technical challenges therein. Rather, the point is that drinking water issues must be situated within the larger framework of land and water dispossession for private industry's accumulation of wealth, and to secure Canada's political and economic sovereignty — all of which is facilitated and legitimated through settler colonial laws. For example, Aboriginal rights have never been factored into water allocation regimes in British Columbia. Meanwhile, there are ongoing disputes over how water figures into various treaties and land-focused legal claims signed between Indigenous peoples and the Crown.

In Treaty 9, the language of water was first introduced in the treaty agreement in the 1929–30 adhesions.[15] Like the earlier clauses on land in the treaty documents, the inclusion of water is specifically to secure and legitimate the Canadian government's access to *Mushkegowuk* territory for settlement, immigration, trade, travel, lumbering, and mining, as well as any other purposes identified by colonial authorities. People of Treaty 9 continue to contest the government's interpretation of the treaty agreement based on Lockean conceptions of property, and the notion that our ancestors actually ceded and surrendered *aski*. Indeed, the very idea of selling kin, our land and water relatives who are legal actors in their own right, does not fit into *Mushkegowuk* political and legal frameworks. Moreover, the people of Treaty 9 have extensively argued that the treaty agreement was corruptly and illegally formalized. Yet the state and industry continue to strategically deploy colonial readings of the treaty agreement, to secure their own political and economic sovereignty.[16]

As such, many people in *Mushkegowuk* territory do not believe that structural colonial dispossession and power relations between Indigenous peoples and the Canadian government can be solved through better regulations and simple investments in water infrastructure. Rather, they understand struggles for water, in all its complexity, as embedded in the historical and ongoing rupture of Indigenous peoples' self-determination, including our political and legal relationships with *nipi*. Specifically, community members such as my *kokom* are increasingly concerned about Indigenous peoples' ability to fulfill our responsibilities as caretakers of *nipi*, amidst the extractive developments that have been and continue to be facilitated and legitimated through colonial Canadian law.

reclaiming kinship responsibilities in and through nipi

In 2011, CLFN community members requested a moratorium on mineral exploratory drilling. The Chief at the time, Arthur Moore, filed a motion in the Ontario Superior Court to seek an order that would stop Zenyatta Ventures, a mining company based out of Thunder Bay, Ontario, from exploratory drilling. Simultaneously, community members protested at Zenyatta's exploration base camp. Meanwhile, members of Marten Falls First Nation blocked the runway at their airport to prevent mining executives from landing in the community. Yet these engagements with the state and embodiments of direct action are just a few of the practices that make up a larger constellation of Indigenous resistance in *Mushkegowuk* territory. The more visible forms of resistance are incited, guided, and sustained through everyday practices of reclamation and resurgence that renew relationships with *nipi*. As my *kokom* says: "You need to go there." She's referring to *Kishiichiwan*.

Over the last several years, Indigenous peoples of *Mushkegowuk* territory have organized community paddles on their waterways. CLFN community members organized a paddle to *Mammamattawa*, a cultural land-based camp that is located at the convergence of the *Kenogami* and *Kabinakagami* rivers. While my *kokom* was not able to physically attend, some of her children and grandchildren were there, such as Arthur, Elizabeth, Luke, and Jacob Moore. They were there continuing the paddling traditions of our ancestors, embodying what my *kokom* and *moshoom* taught them, and taking on leadership roles to teach others.

Meanwhile, community members of Fort Albany First Nation started the *Paquataskamik* Project, or the Albany River Coalition. *Paquataskamik* is a Cree concept that can be roughly translated as the expansive and interconnected ecologies and kinship relations of *Mushkegowuk* territory. As founders of the Coalition say, the concept "reminds us that *Mushkegowuk* land is vast. It's not just the reserve, it's not just the camp (where the project takes place), but an area that ties together family, history, and identity." Like the paddles organized by CLFN community members, the *Paquataskamik* project is aimed at connecting youth with Knowledge Holders and Elders while engaging in land and water-based practices. More than this, one of the aims is to educate community members — and particularly youth — about *Mushkegowuk* water struggles, as they are entangled in unsustainable and illegal developments such as mining.[17] The members of the project do this through clean-up projects on *Kishiichiwan*.

In other instances, *Paquataskamik* has organized community mappings of *Mushkegowuk* place-names and sacred sites along rivers. For example, a group of Elders, adults, and youth travelled from CLFN, down the *Kabinakagami*, *Kenogami*, and *Kishiichiwan* rivers, to Fort Albany First Nation. Throughout their journey, the youth and adults interviewed Elders about the sites and stories along the rivers, and erected signs of *Mushkegowuk* place-names and sacred sites as an act of reclamation. Once the journey was complete, they created a community map that has been widely circulated throughout *Muskegowuk* territory. One version of the map has all of the *Mushkegowuk* place-names written in Cree syllabics.

Through these paddles, community members learn that *Mushkegowuk* peoples have always been water people — people of the river — and always will be. For this reason, we have a responsibility to care for and protect water, so that she can also care for us, and heal us. As the paddles reflect, these responsibilities are learned, transmitted, and embodied through everyday practices that rebuild and reclaim kinship relations with *nipi* and through *nipi*. Simultaneously, these everyday practices re-honour the political and legal authority of water caretakers such as Indigenous women, Elders, and youth, who have been historically relegated from the band council system that continues to shape many governance decisions in *Mushkegowuk* territory.[18] This even applies to the political and legal authority of our non-human kin.

In this way, these are living examples of Indigenous peoples (re)creating their/our own circles of governance that are accountable to *nipi*, as well as kinship relations cultivated through *nipi*. This starts at the most intimate scale, as responsibilities and love for *nipi* are renewed through embodied and storied practice. This ripples outward, as *nipi* reminds us that our kinship responsibilities are not confined to the immediate present, nor to our immediate surroundings. They expand throughout time and beyond the colonial confines of reserves, treaty territories, provinces, and even the geopolitical boundaries of colonial Canada. That is, *nipi* teaches us, through embodied and storied practice, that *Mushkegowuk* kinship relations expand across the spatio-temporal life of our waterways and, thus, our responsibilities do too.

notes

1. *Nipi* is Cree for water.
2. *Kokom* is Cree for grandmother.
3. *Moshoom* is Cree for grandfather.
4. My *kokom* lived in Fort Albany First Nation and my moshoom lived in English River. *Kishiichiwan* translates to "fast water."
5. *Sibi* is Cree for river.
6. My *kokom* attended St. Anne's Indian Residential School in Fort Albany, and my moshoom attended St. Joseph's Indian Boarding School in Fort William Ontario.
7. *Wabaanomakun* translates in English to "white brew." Tea pulozz is Cree slang for *wabaanomakun*. It is tea mixed with fat, either animal fat or lard. My grandparents, like many other *Mushkegowuk*, Anishinaabe, and Oji-Cree peoples of the Treaty 9, used to drink tea pulozz on their trapline during the winter and continue to do so in their homes.
8. *Aski* is a Cree word that encompasses land and water of a territory. It is a concept that expresses the holistic relationship of land and water, and which does not set up a binary between land and water.
9. Personal interview with Angela Moore, June 2016.
10. I use the term Knowledge Holders to denote Indigenous peoples who are knowledgeable authorities, intellectuals, and leaders but are not yet considered Elders.
11. Initial exploratory drilling has also identified deposits of graphite, copper and nickel in the region, with speculations that additional diamond deposits will be discovered.
12. See Vicki Lean's film *After the Last River* (2015) for an extensive examination of the impacts of mining in Treaty 9 territory.
13. Water security and governance within Indigenous communities across Canada continues to be entangled in, and obstructed by, the settler state's purposeful deployment of colonial jurisdictional laws. Specifically, the multiple branches of Canadian government strategically manipulate the multiple levels of jurisdictional law in Canada to evade responsibility and accountability to Indigenous sovereignty.
14. The water advisories range from long-term advisories that have been in place for over a year, to short-term temporary water quality issues on a specific water system. Health Canada identifies three different types of water advisories including: 1) a boil water order; 2) a do not consume order; and 3) a do not use order.
15. Treaty 9, or the James Bay Agreement, was first signed in 1905–06.
16. For a more extensive examination of the Treaty 9 agreement, see Alanis Obomsawin's *Trick or Treaty?* and John Long's *Treaty No. 9: Making the Agreement to Share the Land in Far Northern Ontario in 1905*.
17. I use the term "illegal" here to stress the lack of full and informed consent in most resource extractive negotiations with Indigenous communities in Canada. I also use the term to denote that state and industry continue to strategically mobilize colonial readings of the Treaty 9 agreement.
18. This statement is not to disrespect all of the leaders in *Muskegowuk* territory who tirelessly fight for *Mushkegowuk* sovereignty and well-being through the band council system. Rather, my intention is to stress that the band council system is itself a colonial structure that has historically excluded the political and legal authority of many important Indigenous leaders and decision-makers, along the lines of Eurocentric conceptions of gender, sexuality, and age.

acknowledgements

A deep and heartfelt *meegwetch* (thank you) to my *kokom* and *moshoom*, Angela and Bert Moore, and to members of the Moore family. *Meegwetch* to my mother, Jackie Moore, for comments on an earlier draft of this paper. *Meegwetch* to Jim Ellis for inviting me to contribute to the "Water in the West: Rights of Water/Rights to Water" community forum at the Calgary Institute for the Humanities, and for inviting me to contribute to this book. *Meegwetch* to all Indigenous caretakers and protectors of water — past, present, and future — across our diverse and interconnected waterscapes.

bibliography

Canada, Health. (2017). Drinking Water Advisories: First Nations South of 60. https://www.canada.ca/en/health-canada/topics/health-environment/water-quality-health/drinking-water/advisories-first-nations-south-60.html [Accessed August 2017].

Lean, Vicki. (2015) *After the Last River*.

Long, J. S. (2010). Rupert's Land Record Society Series: *Treaty No. 9: Making the Agreement to Share the Land in Far Northern Ontario in 1905*. Montreal: McGill-Queen's University Press.

Moore, A. (2016). Personal Interview. Constance Lake First Nation.

Obomsawin, A. (2014). *Trick or Treaty?* National Film Board of Canada.

Stevens, S. (2016). Personal Interview. Constance Lake First Nation.

petrography & water

artist's statement & portfolio:

warren cariou

detail: Suncor_ Smoke_ Steel_ Water. V. 1 of 3. 2015.

The image-making technique I call petrography is literally petroleum-photography: the use of tar sands bitumen as a photographic medium to represent the effects of oil extraction upon the land and the people of the Athabasca region. The process, modelled after Nicephore Niepce's famous photographic experiments in 1826, involves spreading a thin coating of bitumen on a metal plate and then creating a contact print after twelve to sixteen hours of sunlight exposure. The images are developed using a mixture of kerosene and lavender oil, which partially dissolves the photosensitive bitumen, leaving an evanescent and highly reflective image that has a distinctive golden colour. Petrography is an attempt to divert petroleum away from its usual role as the fuel of modernity, and to utilize it instead as a way of seeing — a medium that can reveal something to us about our powerful and dangerous attachment to oil. By representing the fragile and often devastated landscapes within the gleaming medium of bitumen itself, I hope to engage viewers in a process of self-reflection about the role of petroleum in their own lives and in the world.

Many petrographs depict the effects of bitumen mining upon the vital freshwater ecosystems of the Athabasca region. Vast areas of wetland have been drained and stripped away by oil companies seeking access to the bitumen deposits underneath. Enormous quantities of fresh water are also used to transport tar sands slurry through surface pipelines, and water is likewise used in the bitumen processing plants to separate the petroleum from sand and other particles. In the extraction process known as Steam-Assisted Gravity Drainage (SAGD), steam is injected deep into the earth to loosen bitumen and transport it to the surface. Bitumen processing plants are also surrounded by gigantic tailings ponds, in which untold millions of litres of toxic wastewater are kept separate from the surrounding boreal ecosystem. Of course, such attempts at containment are necessarily flawed: all tailings ponds inevitably leak, to a greater or lesser extent, and when these leaks occur, the watershed naturally carries much of that tainted water into the nearby Athabasca river, where it affects the health of fish, animals, and humans living downstream. In my 2009 film, *Land of Oil and Water*, I travelled up and down the river, examining some of these effects in detail. I consider petrography as an extension of that work — another attempt to reveal what is being done to the land and water of my home territory in the name of economic development.

My process of petrography is also closely connected to water, because the tar I utilize to make the images is gathered from naturally occurring bitumen deposits on the banks of the Athabasca River, not far from the tar sands mines. However, I source the bitumen from areas that are undisturbed by industry, places where the riverbank remains much the same as it would have been in the eighteenth and nineteenth centuries, when my Métis ancestors would have traveled on the river during the fur trade. For me, it is always a shock to encounter the richness and beauty of the Athabasca's unaltered shores after I have travelled past the excoriated landscapes of the bitumen mines. One might imagine that these natural bitumen deposits would be mostly devoid of life, just like the mine sites are, but in fact the bitumen in its undisturbed state is capable of sustaining a staggering variety of flora and fauna. In these natural riverbank areas, the rainwater that trickles down hillsides of exposed bitumen doesn't seem to be harming the plants below — if anything, it seems to be making the vegetation even more lush than elsewhere. Seeing this verdant landscape each time I gather my bitumen samples, I have come to believe that the toxicity of petroleum is more connected to the extraction process than to an inherent quality of the material itself. It is what humans *do* to the substance that seems to make it dangerous. By the same token, I believe that humans can choose to have a different relationship to petroleum, one that builds from its creative capacity rather than turning it into an addictive and toxic commodity. Petrography is my attempt to negotiate such a relationship by placing myself into intimate proximity with this sticky and smelly substance, working with it to create a different vision of the land, one that I hope resonates with viewers.

I am inspired in my practice by the Indigenous people of the Athabasca region, who long ago developed an alternative relationship with bitumen, using it not as the fuel for an unsustainable lifestyle but instead as a sealant for their birch bark canoes and other vessels that enabled them to travel on the river. When I gather my bitumen now, I often think about those Cree, Dene, and Métis people who searched for this powerful and useful substance in years gone by. Local traditional knowledge about how to find and process the bitumen for use in boats has been lost — or at least I have not been able to learn about it from the Elders and other community members I have met — but nonetheless I take inspiration from the Indigenous harvesters and traditional medicine gatherers I have known, who approach the land with a reverence and an intimate knowledge that is rare and highly valuable in our present age. When I am searching for the places where soft bitumen seeps to the surface, I have to travel the land with a particular attentiveness and knowledge that is only gained by being there, immersed in the landscape. Each time I gather the bitumen, I leave an offering of tobacco or sweetgrass to mark my gratitude and my ongoing relationship to the place where the bitumen comes from.

While most petrographs depict the wounded landscapes of the bitumen mines, I have also begun to create petrograph images that represent the natural beauty of the boreal forest in its undamaged state. I do this as a gesture toward remembering the land as it once was, and also as an acknowledgment of bitumen's own potential to be a generator of new life, rather than solely an agent of toxicity and destruction. By imagining a new role for this substance, I believe we can begin to set a new path for our relationship to the land itself.

Water Treatment Facility on Bank of the Athabasca River. V. 1 of 3. 2015.

Syncrude Tailings Pond Reflections. V. 1 of 3. 2015.

Tailings Pond with High Voltage Girders. V. 1 of 3. 2015.

Suncor and Athabasca River. V. 1 of 3. 2015.

women, water, land
writing from the intersections

helen knott

Water has connected to me in an infinite number of ways. Water has communicated to me in an unspoken language that my grandmother once understood, and it has healed me without asking for anything in return. Water has shown me how to give of myself and nurture those around me unconditionally. Essentially, water has given me my heart for the people.

How do I explain this relationship in this cold language? A language that comes from a people who never understood us the day they brought it in their mouths on ships over 500 years ago. How do I explain it myself when I don't have the language of my grandmothers living on my tongue any longer? The words have been washed away from my family's mouth before I was even born.

I have reclaimed words relating to water.

Choo, *water.* Saghii nachii, *big river*. Tse lingay, *creek*. Mingeh, *lake*.

I may not be fluent in the Dane Zaa language, but it is a language that navigates my blood like the birch bark canoes navigated the river ways. It is a language that is living in my bones. I am still in the process of remembering who I am as a Dane Zaa and Nehiyaw woman whose great grandfathers and grandmothers roamed the land and the waters. My memories are tied to land and water.

During the making of a documentary that I took part in, which focused on the connection between violence against Indigenous land and against Indigenous women, I was asked how to symbolize my healing that has come from the river. I asked what they meant.

"What does it look like? When you are healing by the water? How do we show that on film?" the producer asked me.

I chuckled.

"It literally looks like me sitting down by the river . . . for hours. It looks like tobacco offerings given and prayers said with my toes touching the lip of the water. It's not a razzle dazzle Hollywood moment. In order to learn, you have to be quiet and listen to what the water and the land is telling you," I replied with a hint of amusement in my voice.

I went quiet after that as the thought crossed my mind about what will change if Site C, the proposed mega hydroelectric dam in Northeastern British Columbia, continues to be built. How do I take my pain, tears, and gratitude to a reservoir instead of a free flowing river? I try not to think of such things as it bombards me with an overwhelming grief that I can feel trying to settle in my bones.

■ ■ ■

The relationship that I have with water has caused me to work to protect it. For me "activism" is more about upholding my end of a relationship and my responsibilities. I am fulfilling my role as a Dane Zaa and Nehiyaw woman when I speak out for the water, when I toe a frontline, when I make a tobacco offering, and when I am mindful about how land and water decisions are going to affect all of my relations. I have never ventured into activism out of interest or desire to be a part of something. I actually have struggled with conflict in the past and am shy by nature but fulfilling responsibilities leaves me little room but to stand and to speak. We step into roles out of inherent responsibility.

So I have helped organize at a grassroots level to fight against Site C, BC Hydro's mega hydroelectric dam that could potentially flood eighty-three kilometres of the Peace River Valley. The Valley is full of cultural sites and medicines, is a migratory corridor, and has the best agricultural land north of Quesnel. The Peace River already has two dams built upstream and has slowly been recovering from the methylmercury poisoning it incurred when this happened. Currently you are only able to eat one fish a month from the river and if Site C is to go through, the fish will be inedible altogether for some time as the methylmercury levels will spike again.

I was working and living in Kelowna when the 2014 provincial approval for the dam happened under Premier Christy Clark, and that was when I knew it was time to come home. Prior to that I held a grassroots youth camp where we had Elders come in and reconnect young people back to the territory through stories.

I moved home and in the winter of 2016, along with other First Nation members and local farmers, I peacefully occupied a camp in an old-growth forest that was slated to be logged and turned into a waste dump for acid-generating rock in preparation for construction of the Site C dam. We were located at the historic Rocky Mountain Fort, a place that marked the first solid relationship between settlers and Indigenous peoples within this territory. Each morning we kept fire and our presence there held the gleaming yellow beasts at bay that wanted to clear the tract of land. During that time BC Hydro launched a civil suit against myself and six other individuals but when this threat of a multimillion dollar lawsuit did not stop us they sought out and received an injunction to dismantle the camp. In September 2016 I helped organize and went on a cross-country caravan where we stopped in major cities to talk about Site C en route to attend a Federal Court of Appeal case in Montreal. There have been many communities, people, and organizations, such as Amnesty International, who have come together to make each of these efforts happen to put a stop to Site C. Still, the construction persists in spite of lack of free, prior, and informed consent.

■ ■ ■

There are aspects of how the dam will impact the people that are harder to explain, as they leak into the lives of people and trickle down into their homes, with only traces of its original cause. Impacts of adverse water-related decisions on Indigenous people cannot always be quantified and placed into charts and graphs. Anderson, Clow, and Haworth-Brockman (2013) state that "water quality issues can threaten spiritual and cultural well-being at the same time that they endanger physical health, local and household economies, and the environment" (12). The multifaceted impacts are lived realities for Indigenous people who experience relationships with the water, land, medicines, and animals within their territories. We are still trying to gain traction for our long-lived truths to be accepted in a world that, dominated by Western knowledge, relies heavily on what can be quantified.

How do you place a number on prayer?

How do you chart genuine connection?

What price do you place on healing?

How do you create a statistic to show how crucial land is in regards to identity formation as it is the gateway to blood memory and acts as storyteller?

At what point do we accept that other ways of being and knowing are valid?

Are we really still having this conversation?

It has been 150 years and Canada still has not learned how to accept and respect Indigenous ways of being and knowing.

A prime example of this would be in the winter of 2016 when Amnesty International released their report "Out of Sight, Out of Mind," which examined the connection, in the traditional territory I am from, between violation of Indigenous lands and violence against Indigenous women and girls. Three of us women stood at the press conference for the release of the report and gave our own personal testimonies that were included in the report alongside many other identified and anonymous testimonies. A reporter raised his hand shortly after we were finished and asked for "hard facts," and then pushed a few more times for numbers and statistics that were not present in the report. He was after the quantifiable and tried to kick the legs out from under the power of oral testimony. We are an oral people; all we have is our stories, and the difficult ones are ones that we tell even when it is hard, because we have the audacity to hope for change.

■ ■ ■

Change. It's a funny feeling, to empty yourself into efforts towards the elusive. It's an even funnier feeling when those efforts have been towards elusive justice. Justice. Not materials, not dreams, not a larger bank statement, not trinkets, nor personal ventures, but Justice. When you believe that the right will always prevail and time and time again it does not happen, you feel a twisting in your chest. A wrestling of the heart trying not to give up on itself.

To be Indigenous and in pursuit of the preservation of land, water, and inherent rights that belong to those who come after you because of those who became before you – while simultaneously experiencing the elusiveness of justice yourself – is the funniest thing of all. Not funny in the satirical dark, humour kind of way – the way through which I have learned to laugh at most of life's maladies and afflictions. Funny in an existentialist "what the fuck is happening, how do I relate to this world in which I am existing" kind of way.

Why? Because you can't just simply exist in such elusive moments. Indigenous activists who are expected to engage with media often know that they will be interpreted. We learn to monitor ourselves and our responses to these moments in order to be properly consumed by white audiences. Present your grief in a palatable way, don't throw wild accusations, keep your head, and make sure your pain is tasteful – always balanced with possibility. Perhaps I say "we" too presumptuously but I know that this is the pressure I have often felt, from multiple avenues.

I've never been fond of censoring myself, but I have an inner dialogue that says don't be too sorrowful because it will make "them" think that they are winning. Don't sound defeated, tell people to keep hope and to become louder than ever. Be that warrior, be the fierceness you have witnessed in many other land protectors, be the fire that threatens to consume, be the thunder that makes its presence known. But be honest, always be honest. Just be an honest version of inspiration that is available for public consumption. Media engagement as an Indigenous activist can be a brutal battering ram and if you are not good at placing boundaries and staying out of the comment section, where racism breeds, it is ultimately traumatizing.

The experiences of removal from land and at the same time losing the land itself to projects is also traumatizing. Dispossession in a dual sense. After the dismantling of our winter camp I was despondent, had a short attention span, kept to myself, became depressed and disinterested for months before I figured out I was dealing with trauma from the loss of the camp. I've done intergenerational trauma workshops as a social worker within the communities in the north so I understand trauma but even I did not catch this right away. When I did catch it I was crying in a coffee shop as a journalist ripped open wounds for examination. I have worked to heal myself from the historical traumas my family and people have endured and passed on and I have to work continually to heal from the present-day colonial lacerations. We need to be mindful that colonial goals of assimilation and inflicting trauma on Indigenous people are still very real and present to this day.

At the time of writing, the new NDP and Green Party provincial government has sent Site C for review and in the coming months we will see if the dam will finally be stopped. There is hope, however fragile. There is hope. We who stand on frontlines cannot live without it, and our children and grandchildren cannot afford for us to lose it, either.

bibliography

Anderson, Kim, Barbara Clow, and Margaret Haworth-Brockman. (2013). "Carriers of Water: Aboriginal Women's Experiences, Relationships, and Reflections." Journal of Cleaner Production 60: 11–17.

Photographs courtesy of Helen Knott

Alexis First Nations: Lac Ste. Anne Shoreline, 2011, Digital print on rag paper, 32 X 23", Collection of the artist.

tanya harnett: the poetics & politics of scarred/sacred water

nancy tousley

Paul First Nation: 2005 Wabamum Clean Up Site of a 700,000 Litre Oil Spill, 2011, Digital print on rag paper, 32 X 23", Collection of the artist.

For two weeks during the summer of 2011, Tanya Harnett travelled to four reserves in the Treaty 6 territory of northern and north-central Alberta, and to the un-ceded traditional lands of the Lubicon Lake Nation, where there are problems with the water. The most highly publicized damage was caused by the CN train derailment, in 2005, that spilled 1.3 million litres of bunker oil and hazardous chemicals into Lake Wabamun on the Paul Band reserve. Harnett knew of other damaged water from news stories or by word of mouth. In addition to Lake Wabamun, she drove to Cold Lake, the Driftpile River, Lubicon Lake, and Lac Ste. Anne. The size of the damaged sites varied, from lakes to a river and a spring, but the water and the people — Cree, Chipewyan, Nakota Sioux, and Nakoda — who live on these reserves had suffered and still were suffering significant distress. She went there to photograph these waters because the damage done to them is largely unknown beyond First Nations communities. To bear witness and to give testimony was Harnett's way of framing the question that formed the impetus behind the artwork she was undertaking: what is the Indigenous perspective on landscape?

"Scarred/Sacred Water" is her answer, a series of six un-manipulated, digital inkjet prints that Harnett[1] produced from photographs she made on the reserves. In each of the six images, damage to the water is represented viscerally as a red or bloody wound on an injured body. It is her view, borne out by statistics and reports,[2] that the crisis in the health of water on reserves in Alberta, and in Canada as a whole, is the biggest issue concerning Indigenous people in their relationship to the land. This complex issue, or rather the issues, are not only physical, political, social, and economic, but also and most importantly, spiritual. Water is the sacred giver of life.

When we talked, Harnett — who is Nakota and from the Carry the Kettle Nation, located in southern Saskatchewan on Treaty 4 territory — outlined the protocols of her visits. Although she is also Indigenous and Albertan, she is not a person from the Treaty 6 territory where she wanted to photograph. At each reserve, she knew or sought out a band member and asked for permission to be there. She asked where the water was damaged and to be taken to the site and allowed to photograph it, using this analogy to explain: she could not just come into my backyard unannounced and start taking photographs. Harnett or her local contact released concentrated red food colouring at each site to mark and reveal the damage, which was not apparent to the eye. She then photographed the marked site. The red food colouring was ephemeral and would disappear. The six photographs that make up "Scarred/Sacred Water" would circulate publicly in the world beyond the reserves, making visible what once was invisible and giving testimony to what she saw.

The work was commissioned by the Alberta Foundation for the Arts Travelling Exhibition Program, known as TREX, for the exhibition entitled "Our Wilderness is Wisdom."[3] Harnett envisioned "Scarred/Sacred Water" as an antidote to the grand vistas conceived in the sublime or picturesque European Romantic traditions that idealize and distance the landscape, and affirm the identity and power of a nation (or province). "Scarred/Sacred Water" presents viewers with the everyday poetics and politics of the local, seen up close, candidly, and in detail. The series speaks to the concrete realities of people on the reserves, whose plight is part of a longstanding water crisis that exists nationally. The photographic images are beautiful — atmospheric in colour, and delicately nuanced in their clearly articulated surfaces. Nevertheless, these are images that have been constructed to provoke questions. In Harnett's photographs, blood red is a punctum in the Barthesian sense: a detail that pricks or bruises, or wounds.

What has happened here? Even a child looking at these photographs would see that something has gone wrong. Because the TREX program circulates exhibitions to schools, museums, and libraries province-wide, Harnett knew that an important portion of the audience would be children. She made the work for them, she says. But who could *not* see that something was wrong? Harnett enlisted Syncrude, the exhibition's sponsor and one of the largest producers of crude oil from the Alberta oil sands, to educate people, everywhere this work is seen, about the deleterious effects of industry on the environment and, specifically, the Indigenous land of reserves. It was a nimble, non-confrontational political move to accept the commission and then make the work she chose to make.

But, as the title of the series suggests, "Scarred/Sacred Water" voices layered concerns that go beyond the ecological to the cultural and spiritual. It was the poetics of her images that engaged me first. Or more accurately, what I read as the poetics of the work. When I began to look closely at "Scarred/Sacred Water," my Western-educated thought processes took me directly to landscape and metaphor. Until one morning in the shower: with water pouring over me, I realized that although she was well aware of it, landscape as metaphor or text was not what Harnett was about. To echo Lacanian terms, I was absorbed by the Imaginary, while Harnett was addressing the Real, the land. I read metaphor; she saw actuality. But then the whole idea of landscape is a construction that was brought to North America by colonizing settler culture. W.J.T. Mitchell pulls the veil from the scenic view: "Landscape, we suggest, doesn't merely signify or symbolize power relations; it is an instrument of cultural power, perhaps even an agent of power that is (or frequently represents itself as) independent of human intentions. Landscape

Cold Lake First Nations: Damaged Spring at Blue Berry Point, 2011, Digital print on rag paper, 32 X 23″, Collection of the artist.

Lubicon Lake First Nations: Warning Oil Pipeline, 2011, Digital print on rag paper, 32 X 23", Collection of the artist.

as a cultural medium thus has a double role with respect to something like ideology: it naturalizes a cultural and social construction, representing an artificial world as if it were simply given and inevitable, and it also makes that representation operational by interpellating its beholder in some more or less determinate relation to its givenness to sight and site."[4]

How one reads *Scarred/Sacred Water*, then, can depend not only on one's point of view but, more fundamentally, on how one understands the nature of the world. Central to the artist's work is the question: is nature animate or inanimate? Indigenous belief answers that it is animate. A settler perspective raises other questions. If nature is animate, do Harnett's photographs represent landscape or a body? Is the scarring of the title a metaphor or something more? Is water a resource or a sacred element? In terms of art, are Harnett's photographs documents or constructions? Whichever way they are received they are agents of narrative, as photographs often are. On the one hand, they are visual news and, on the other, visual narratives of distinctive character.

Cleaving to Indigenous cosmology, Harnett works against the European construction of landscape and the connotations of property and commodity that attach to it, which Mitchell deftly exposes. In opposition to notions of landscape, Harnett addresses the land directly. In her artist's statement in the interpretive guide to "Our Wilderness is Wisdom," she writes that when she made the photographs she was "listening to the land." "Although this idea of a land 'speaking' might sound odd to some, the notion and the understanding of the concept is core to native spirituality. Absolutely everything has a spirit, energy, a resonance, a meaning and everything is connected. These connections might be abstract philosophies, but simply put, everything touches everything, like an invisible web. We are, individually, a meek part of the metanarrative."[5] In another context, Loretta Todd writes: "As the old people have said: *The land is the culture*."[6]

Harnett works in and through the belief system of Indigenous culture, even as she employs the technology and analytical tools of Western culture and thought. Her images of "Scarred/Sacred Water" are dialectical images, which synthesize to create a form of storytelling, whose visual rhetoric can be parsed. She frames the *mise-en-scène* of each photograph in the vertical portrait format, choosing it instead of the expected horizontal format associated with landscape. The phrase "body of water" in English transfigures water with a metaphor. Harnett goes beyond metaphor. She envisions, through embodiment, a wounded and bloody corpus to substantiate an aspect of Indigenous cosmology. The red connotes the body as well as the wound. These Indigenous bodies of water are presented as living characters with distinctive personalities, much as natural entities are represented as characters in Indigenous stories and oral histories. All of the waters that Harnett has photographed, in fact, have stories attached to them by their communities, which tell and perpetuate their histories in memory and over time. In Harnett's photographic images, it is the waters who do the telling by the showing of their wounds.[7]

These waters cannot speak literally, of course. Harnett gives voice to their characters in the way that she has conceived and photographed their images. First, there is the action that marks the site of the wound with red colouring. The photograph is not a constructed image so much as it is a document of this result of the action. This is also where the visual poetics of metaphor operates: the colour red is a visual metaphor for blood, perhaps universally. Moreover, the releasing of red colouring into the sites is both a marker and a performance akin to a speech act, an utterance: "Here it is." And, more than this, it is an example of the performative, which "constitutes the performance of the specified act by virtue of its utterance."[8] It performs in the here and now to make manifest the wound, which formerly was invisible, while in the same breath it invokes the cultural memory of the past in which the land was thriving and its health was not in question. Realizing what in the vast space of time are recent occurrences, it calls forth the memory of the health and suffering of the land and, because they are connected and inseparable, the memory of the health and suffering of the people and their culture.

This is the power of stories and what they can do, what images can do when their visual rhetoric is parsed. Harnett is an artist and activist, a visual storyteller who honours the oral tradition while bridging cultures through the making of dialectical images — translating, retelling, and creating new stories as she goes. Hers is the work, equally, of poetics and politics.

Driftpile First Nation: Driftpile River from Swan Hills, 2011, Digital print on rag paper, 32 X 23", Collection of the artist.

Lubicon Lake Indian Nation: Damaged Creek and Land Access, 2011, Digital print on rag paper, 32 X 23", Collection of the artist.

notes

1. Tanya Harnett is a well-known multidisciplinary artist, whose work has been shown here and abroad. She is an associate professor at the University of Alberta, where she teaches by cross-appointment in the Faculty of Native Studies and the Department of Art and Design. Ideas and observations attributed to her in this text are derived from two interviews with the artist (recorded on Sept. 6 and Oct. 6, 2017), unless otherwise noted.

2. CBC News investigation in 2015 found that the condition of water on Canadian reserves was comparable to that of Third World countries. "The longest running water advisory is in the Neskantaga First Nation in Ontario, where residents have been boiling their water for twenty years. Nazko First Nation, Alexis Creek First Nation and Lake Babine, all in British Columbia, are next on the list with water problems spanning sixteen years. Between 2004 and 2014, 93 per cent of all First Nations in Saskatchewan and New Brunswick reported at least one water advisory in their communities. Alberta is close behind at 87 per cent." http://www.cbc.ca/news/canada/manitoba/bad-water-third-world-conditions-on-first-nations-in-canada-1.3269500.

   A 2017 report from the David Suzuki Foundation and The Council of Canadians, making the same comparison, found conditions largely unchanged. https://globalnews.ca/news/3238948/first-nations-drinking-water-crisis-liberals-promise/.

3. "Our Wilderness is Wisdom," curated by Heather Shillinglaw, also included the work of Alex Janvier and Curtis Johnson. The exhibition was organized by the Art Gallery of Alberta for the Alberta Foundation for the Arts Travelling Exhibition Program and sponsored by Syncrude Canada. It travelled throughout Alberta in 2012 to schools, libraries, museums, health care centres, and other community facilities.

4. W.J.T. Mitchell, "Introduction," in *Landscape and Power*, 2nd ed., ed. W.J.T. Mitchell (Chicago: University of Chicago Press, 2002), 1–2.

5. http://www.trexprogramsoutheast.ca/files/2012/10/Our-wilderness-is-wisdom...pdf.

6. Loretta Todd, "Yuxweluptun: A Philosophy of History," in *Beyond Wilderness: The Group of Seven, Canadian Identity, and Contemporary Art*, John O'Brian and Peter White (eds.) (Montreal and Kingston: McGill-Queens University Press, 2007), 345.

7. The Whanganui River in New Zealand and subsequently two rivers in India were granted personhood under the law in 2017, recognizing them as living entities. https://www.theguardian.com/world/2017/mar/16/new-zealand-river-granted-same-legal-rights-as-human-being.

8. Mieke Bal, *Travelling Concepts in the Humanities: A Rough Guide* (Toronto, Buffalo, London: University of Toronto Press, 2002), 174.

Bighorn by Stephen Legault

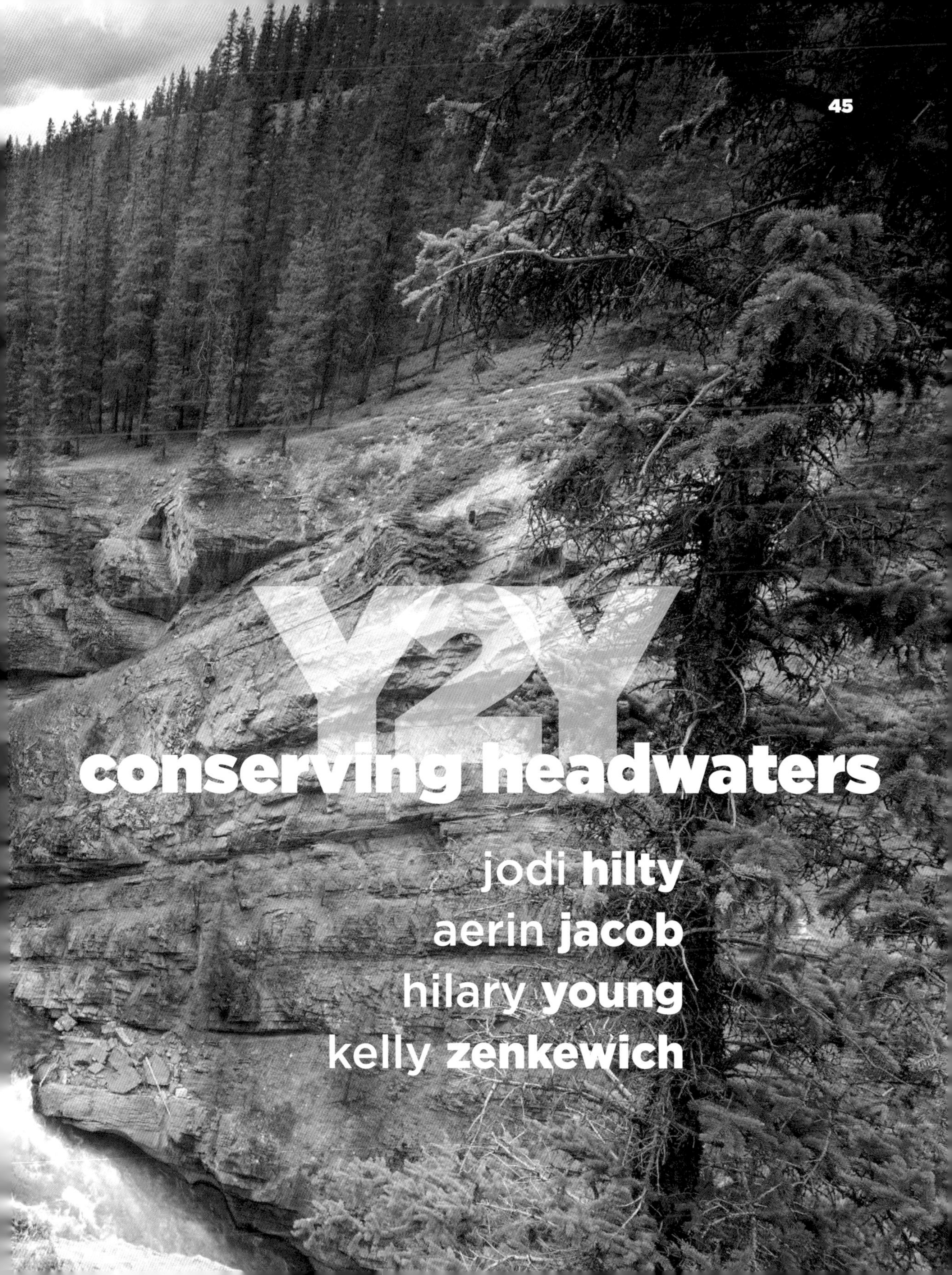

Y2Y conserving headwaters

jodi **hilty**
aerin **jacob**
hilary **young**
kelly **zenkewich**

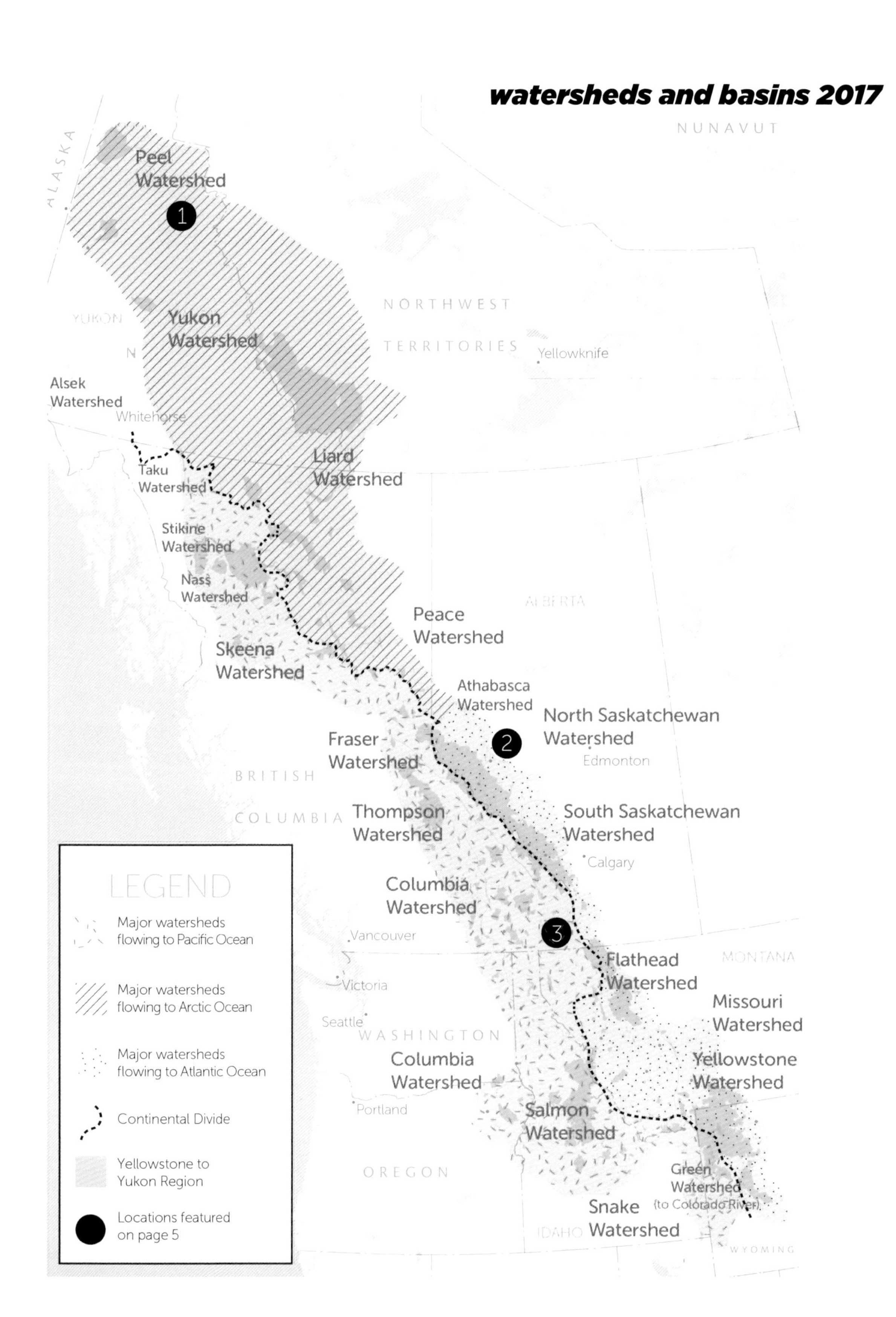

watersheds and basins 2017
NUNAVUT
ALASKA
Peel Watershed
1
YUKON
Yukon Watershed
NORTHWEST TERRITORIES
Yellowknife
Alsek Watershed
Whitehorse
Liard Watershed
Taku Watershed
Stikine Watershed
Nass Watershed
ALBERTA
Peace Watershed
Skeena Watershed
Athabasca Watershed
North Saskatchewan Watershed
2
Edmonton
Fraser Watershed
BRITISH COLUMBIA
Thompson Watershed
South Saskatchewan Watershed
Calgary
Columbia Watershed
3
Vancouver
Flathead Watershed
MONTANA
Victoria
Missouri Watershed
Seattle
WASHINGTON
Columbia Watershed
Yellowstone Watershed
Portland
Salmon Watershed
OREGON
Green Watershed (to Colorado River)
Snake Watershed
IDAHO
WYOMING
LEGEND
Major watersheds flowing to Pacific Ocean
Major watersheds flowing to Arctic Ocean
Major watersheds flowing to Atlantic Ocean
Continental Divide
Yellowstone to Yukon Region
Locations featured on page 5

The lure of a rushing river, beginning its journey high in the mountains, is undeniable. Instantly enchanting, cold, clear waters flowing from mountain-top glaciers are appealing not just for their looks, but for their life-giving properties. Healthy rivers signal functioning headwaters and healthy watersheds. No matter where you live, you are part of a watershed. Falling in the mountains and hills as rain and snow, or melting from glaciers, water drains into streams, rivers, and wetlands and eventually finds its way to your tap. In the North American West, water filling your glass could come from a watershed in the Yellowstone to Yukon region. This area contains major water basins feeding thirteen major rivers, ultimately providing clean drinking water to 15 million people. Live in Vancouver, Calgary, Missoula, or even as far south as California? You could be one of them (see map on facing page).

Beyond their beauty and water supply, rivers supply water for thousands of species of plants and animals, help regulate climate, and cycle nutrients feeding our forests. Given their importance for biodiversity, as well as recreational, economic, and cultural significance for people, it is no surprise that conserving healthy water systems is paramount in Y2Y's work. As an organization dedicated to connecting and protecting the Yellowstone to Yukon region so that people and nature can thrive, Y2Y works to conserve headwaters. For example, our engagement in Nahanni National Park Reserve in the Northwest Territories with First Nations, the Canadian Parks and Wilderness Society, and the federal government resulted in a sevenfold expansion of the park in 2009, protecting the larger Nahanni watershed rather than just the narrow corridor along the river.

Likewise, based on strong science, we led a campaign with partners from other non-governmental organizations and local community groups to increase protection of Alberta's Castle watershed. In 2017, after several decades of work, the Alberta government has delineated new protected areas within the watershed and is phasing out the use of off-road vehicles. The Castle parks are important for connectivity at the Y2Y scale, and form part of Alberta's headwaters region, providing 90 per cent of the province's drinking water but making up only 10 per cent of its land-base. These parks in particular are the source of 30 per cent of the water in the Oldman River, which serves millions of people downstream as it flows through communities such as Fort Macleod and Lethbridge before merging with the Bow River.

The emerging science on climate change shows it will be even more important to conserve headwaters to secure healthy futures for people and wildlife. The spectrum of biological and physical impacts from climate change is immense. In the Y2Y region, the current shrinking and future loss of glaciers will mean that during the hotter summer months, some sources of water may disappear altogether. In the central Canadian Rocky Mountains, glacial coverage has shrunk more than 25 per cent since the mid-1800s. The region also has less snow overall, earlier peak stream flows, and warmer summer and winter temperatures.

To protect biodiversity and maintain water resources, climate scientists strongly recommend expanding and connecting protected areas. This is absolutely crucial in water systems with less predictable, larger swings between drought and flooding. Conserving entire riparian systems is critical, as the areas next to flowing rivers are especially important for biodiversity. New research, which Y2Y contributed to, suggests that gravel-bed flood plains drive ecological processes and interactions far beyond rivers' margins. As such, maintaining functional river systems is an important priority for Y2Y.

Recognizing the importance of watershed conservation in the Y2Y region during this time of climate change, our priorities are twofold. First, we use science to identify watersheds that will be robust to climate change in the long term. For instance, based on climate change projections and modeling, Y2Y is working with Wildsight, a regional conservation group in southeast British Columbia, and other partners to increase protection for the Columbia River headwaters. A large reason for this work is that climate projections suggest that this region is likely to be robust to climate change and serve as climate refugia. Second, Y2Y is working to protect more than 5,000 km^2 of the Bighorn backcountry, an area in central Alberta sandwiched between Banff and Jasper National Parks that makes up the headwaters for nearly 90 per cent of Edmonton's drinking water (Figure 2). Recent research by the Wildlife Conservation Society shows how important this region is for wildlife, fish, and water conservation. This type of work – using science to inform conservation priorities and action – continues to be a top priority across the Y2Y region, including the Peel, Peace, and Snake River watersheds.

Through its partnership approach to conservation, in the first twenty years of its existence Y2Y has significantly increased protected areas across the Y2Y region from 11 to 21 per cent.

Nevertheless, a special focus on water conservation is necessary if we are to succeed in our long-term vision. Such a focus helps us to identify watersheds particularly at risk and ensure the critical ecological processes related to water are maintained or restored for both people and nature.

the elbow river

watershed partnership

flora giesbrecht

about the elbow

Rising in the eastern ranges of Alberta's Rocky Mountains, the Elbow River is small, yet significant. It not only supports the ecosystems within its 1,200 km^2 watershed, but contributes to the lives of the one in six Albertans who drink from its waters. From its mountain headwaters, the Elbow flows through the foothills, past the communities of Bragg Creek and the Town of Redwood Meadows, through the agriculture and grasslands of Springbank, and into the City of Calgary, to enter its Glenmore Reservoir. From there, it meanders through urban communities to join the Bow River at Fort Calgary.

The ERWP provides a forum for learning about watershed management and the land-water connection. Using local knowledge and scientific expertise, it promotes watershed management improvement through collaborative, targeted, and cost-effective actions by citizens and stakeholders. We are surely "working together for a healthy Elbow River watershed."

goals

Encourage individuals and communities to take responsibility for protecting and enhancing water quality and quantity in the Elbow River watershed

Encourage the use of new technologies for water conservation

Encourage best water-management and land-use practices

Support cooperation, coordination, and knowledge sharing among stakeholders

Minimize the negative impacts of land uses on water quality and quantity

Increase awareness and understanding of the watershed

achieving our goals

Achieving our goals requires the collaborative efforts of groups (private sector, government, and public interest) active in the Elbow watershed. The Partnership provides a forum to focus and coordinate watershed management efforts. Actions necessary to protect or maintain the watershed are identified, and the ERWP works cooperatively to implement those actions with appropriate organizations or agencies.

The Partnership acknowledges that individual groups are working and will continue to work towards their individual objectives. The Partnership is not a regulatory group and does not seek the authority to compel action. Member organizations can be expected to fulfil externally mandated (i.e., Legislated or Codes of Practice) responsibilities but it is not the role of the Partnership to ensure that they do so. Agencies with regulatory authority retain that responsibility and there is no expectation that any of that authority will be delegated to the Partnership.

guiding principles

Environmentally appropriate management practices are cost effective

Effective environmental and watershed stewardship practices come from a forum where all stakeholders can participate and contribute

Our approach is open, inclusive, and purpose-driven

Stakeholders are encouraged to contribute their expertise in the process

Stakeholders work in cooperation with one another toward achieving our goals

Previous spread & right: photos by Robert Lee.

We respect and value the opinions of others

The process is flexible to allow for change

Membership is voluntary

We follow the best available scientific understanding of issues, cause and effect relationships, and best management practices

Decision-making in all committees is by consensus

The ERWP has a variety of active projects each year, including our unique flagship partnership program, the Freshwater Field Study. Together with Alberta Environment and Parks, Kananaskis Region, since the program began in 2005 we have educated over 15,000 students from grades 8, 9, and 11 about the headwaters of the Elbow.

some other ERWP projects include:

Elbow River Watershed Management Plan update — this will be an opportunity for broad stakeholder engagement in the revision of our 2009 plan

Riparian Restoration — using natural processes and volunteer labour, we are restoring damaged areas along creeks and wetlands.

Communications — via our online newsletter and other social media, public outreach, presentations, and mini-symposia

New Watershed Interpretive Trail — an interpretive signs loop in West Bragg Creek

Water Quality Monitoring (based on Citizen Science) — this annual program has operated for twelve seasons to date

Other watershed topics for discussion and education include: proposed flood mitigation projects (e.g., the Springbank Off-Stream Reservoir), the Southwest Ring Road construction, wetlands studies, land uses in the watershed (e.g., agriculture, forestry, off-highway vehicles, residential development, and stakeholder initiatives such as The City of Calgary Headwaters Protection Strategy and The City of Calgary Riparian Action Program).

The Elbow's watershed falls within several jurisdictions, including Kananaskis Country, Rocky View County, the Tsuut'ina First Nation and the City of Calgary. The Calgary-based Elbow River Watershed Partnership (ERWP), a non-profit organization formed in 2004, has the responsibility of education and promotion of sustainable management of the Elbow and its watershed.

This image depicts the Elbow Watershed and the jurisdictional boundaries

Please visit our website **www.erwp.org** for more information about the Elbow River and the Elbow River Watershed Partnership.

confluence:

artist's statement & portfolio

leslie sweder

Living in the city, I have observed the tendency of our own urbanity to perpetuate a chronic disconnect from the land of which we are a part. This sense of disconnect drives my explorations of the landscape that has formed me, and my place within it. Out of a discordant situation I have developed a multidisciplinary practice that includes public interventions, painting, collaborative drawing, and photography. These practices are direct responses to and/or recordings of my surrounding environment, and each one implicitly feeds the others. As a whole, my practice is always mindful of people and place and the ongoing conversation between the two. To me, landscape is not what we see from afar. It is what we experience through all of our senses. It is what we are.

Confluence evolved out of this desire to immerse myself in my immediate landscape, to know intimately the movement of energy that surrounds and affects me. In order to know the pulse of our city and the land it grew out of, I chose to focus on its heart, the confluence of the Bow and Elbow Rivers. It is here I document the smaller movements of these rivers through *plein-air* studies (*Notes on Current*) and create photographs of the detritus along their shorelines (*The Things We've Left Behind*). Lastly, I take home objects whose original purpose is unknown to me, or which no longer seem to have a purpose. I study these objects through drawings in the studio (*The Dispossessed*). In *Notes On Current*, I draw on the traditional history of Canadian *plein-air* landscape painting (particularly the small-scale studies of Tom Thomson), the exquisite stillness of Agnes Martin, and the abstract landscapes of Gerhard Richter. I also pay heed to the thinking of artists such as Otto Rogers who believed that "one of the distinct advantages of painting as a discipline is its static nature, its symbolic representation of movement by means of stillness." As a whole, my practice is an investigation into the resiliency of nature and the fragility of life. The paintings are meditations on grace, transition, the linearity of flow, and the eternal cycles of nature. When the paintings are juxtaposed with the photographs and the drawings, which document the smaller visible traces of the human footprint, they further explore our reciprocal relationship with the river. When I look through the water to see the earth beneath it, it reflects my image up to the sky. Yet I am not the least bit interested in my own reflection. I am interested in the river as that liminal space that connects the earth and the sky. It is where we come to understand that what appears solid is empty and what appears empty is not. For this reason, the paintings become studies in space while exploring landscape and abstraction as related concepts. I do not interact with the river directly nor am I attempting to affect it. I allow it to affect me, to show me its nature and to teach me.

The Bow River as it runs today is an estimated 13,000 years old. This fact alone is humbling. I can't fully imagine what 13,000 years means, but I do feel the weight of my own years, and try to imagine all the changes and all that has not changed during this time. I focus on documenting what the water shows me during the time I sit on its bank. Although the river is a continual flow, it is different in every moment. It is never the same water, the same form, colour, or the same energy. You will not see what I paint in a single moment on the river — not as you would in a photograph. These works capture a gradual change in light and movement, including the life moving through it and over it. In this manner each painting becomes a study in time. I study the current of the river, its light, its sound, the smell, and the touch of the cool water gliding over my skin. This is a sensual practice and I use it to mine the many different expressions of the river. In this manner, I document the river's rhythm, movement, and energy. I appeal to the river's natural intelligence, in search of my own, but the river remains elusive. I am never successful in capturing it and somehow this keeps me returning.

It is from the river that I am learning how to move through the world with grace, how to be both soft and strong, how to negotiate conflicting energies, how to absorb new elements while letting things go when they do not serve me. When I look back on these words I have written, they sound lofty, grand — but then again, the river is grand. I am merely its student.

girl on a rock, SE end of st patrick's island bridge
last light, 2017
oil on birch panel
8 x 10″

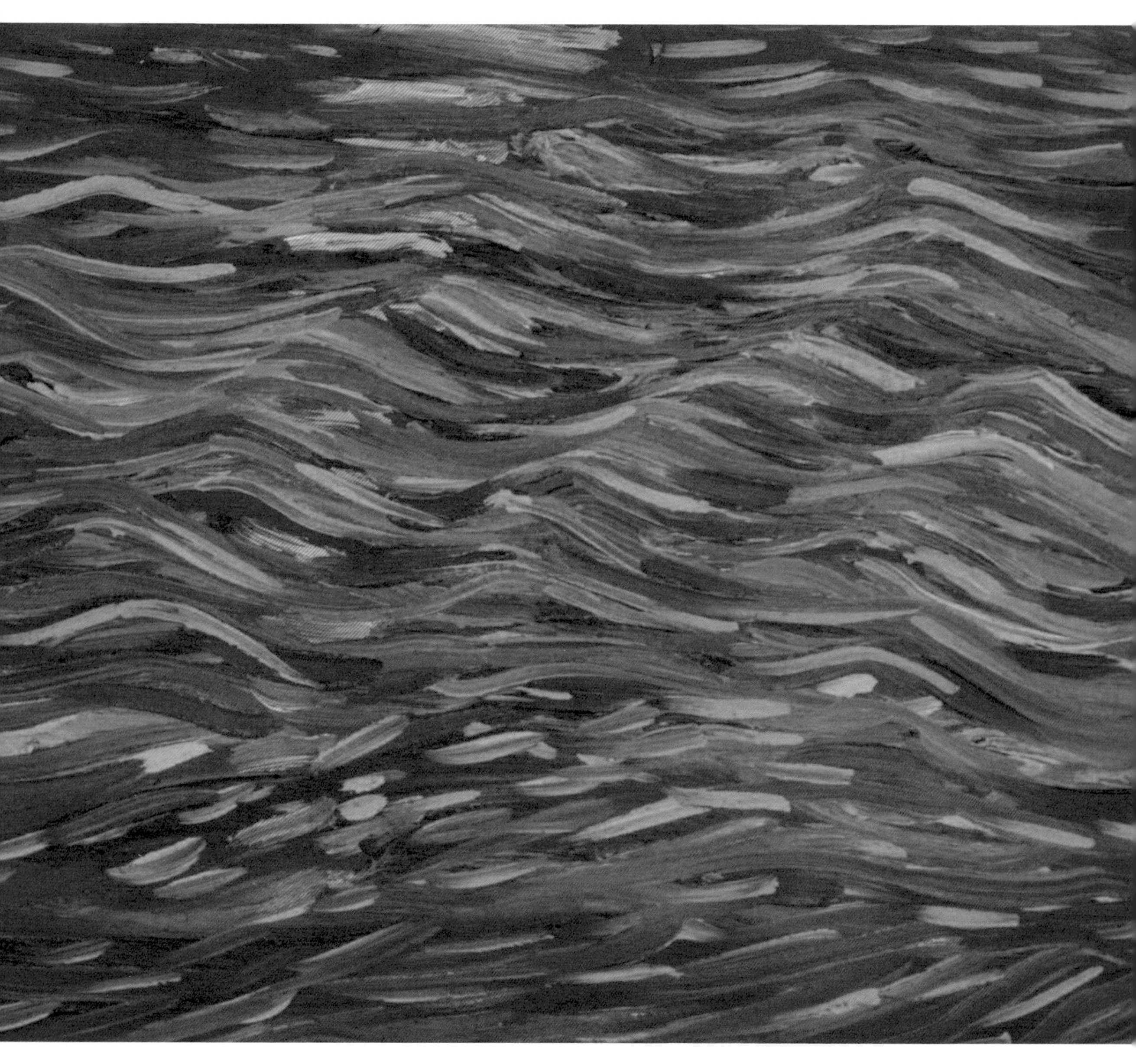

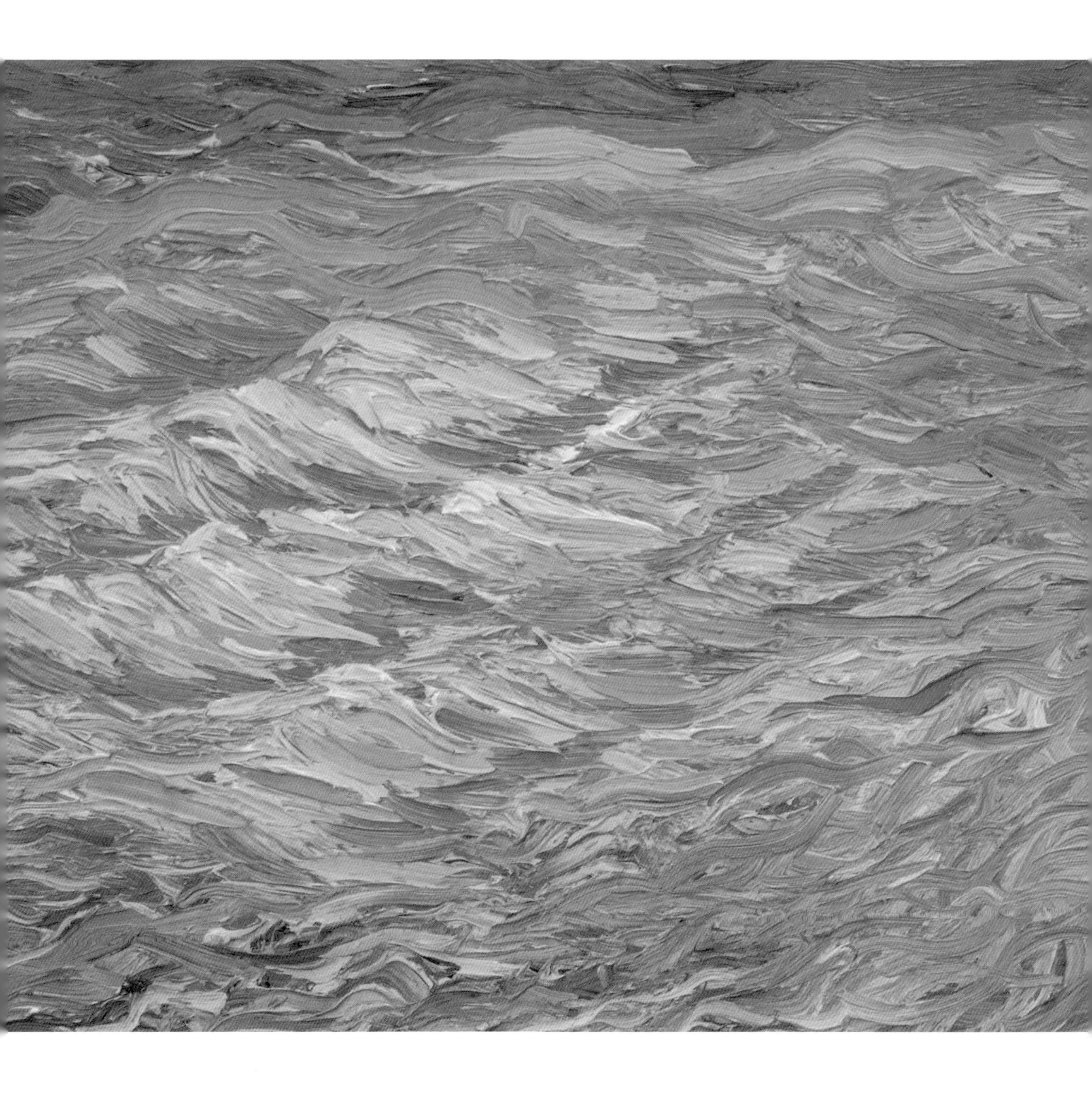

spring run off, SW end of st patrick's bridge
last light, june 20, 2017
oil on birch panel
8 x 10"

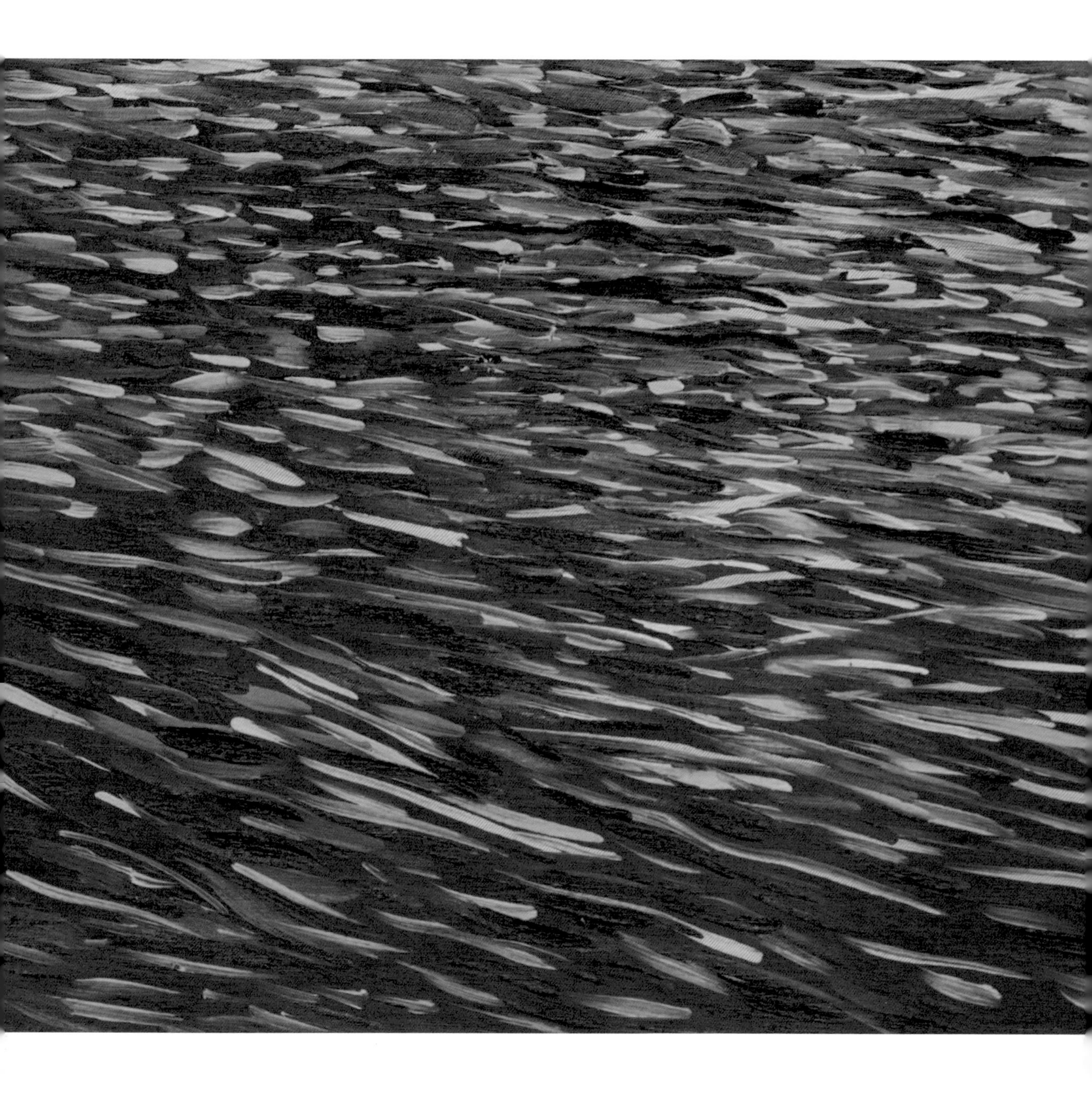

SW corner of confluence
last light, august 17, 2017
oil on birch panel
8 x 10″

new zoo bridge construction
last light, july 8, 2017
oil on birch panel
8 x 10″

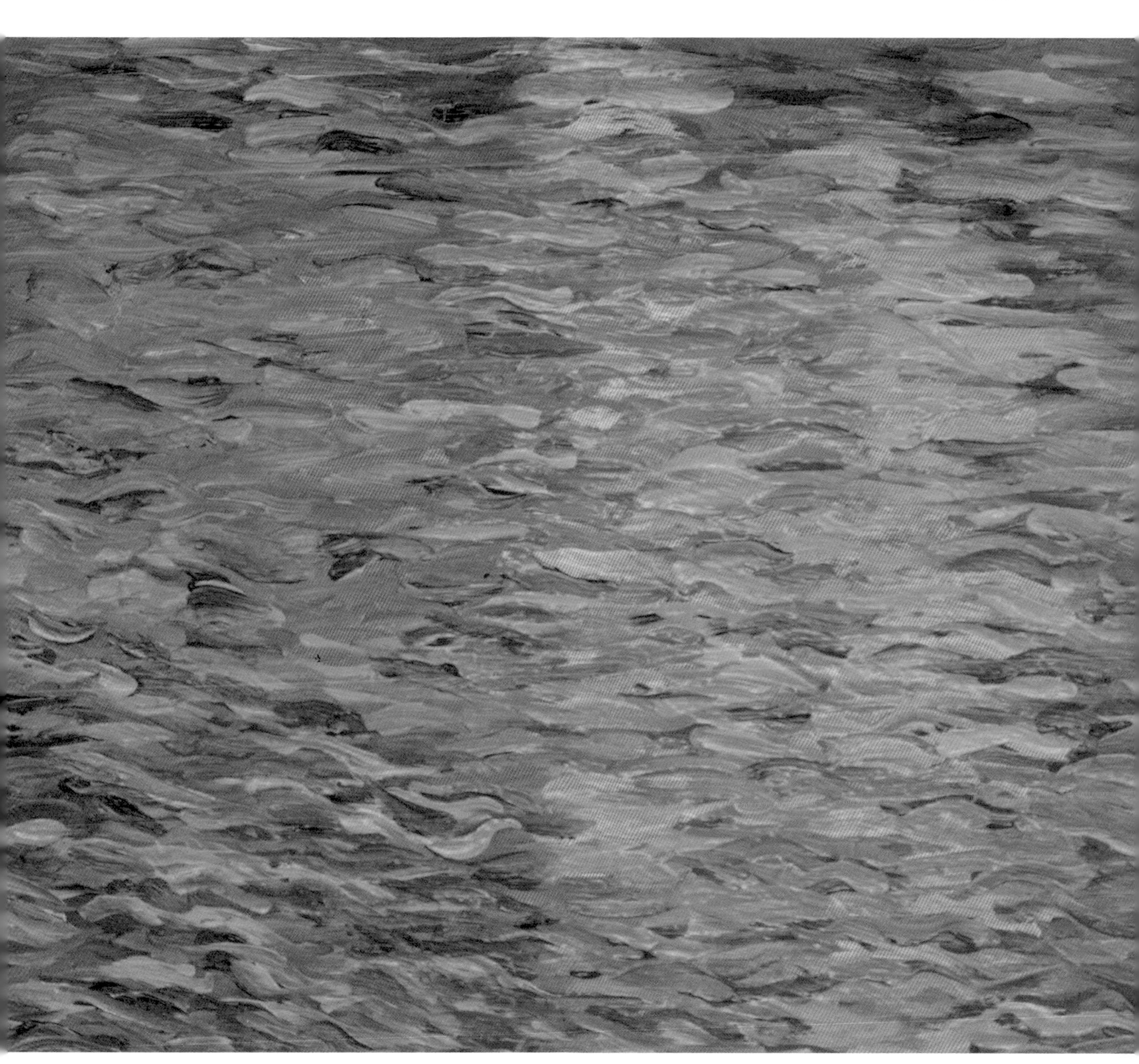

the brown apartment block, SE end of st patrick's bridge
last light, july 14, 2017
oil on birch panel
8 x 10″

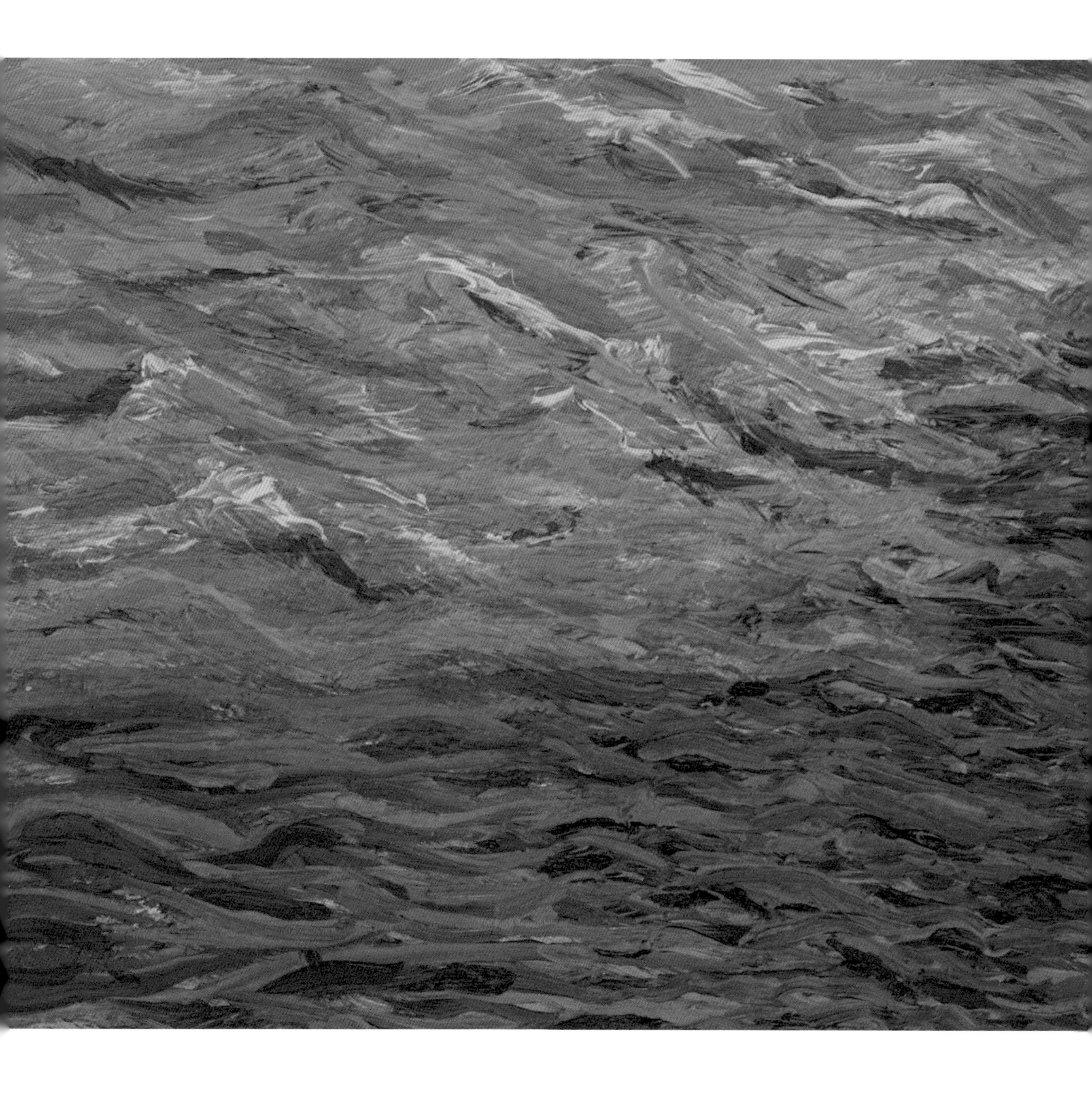

SW corner of confluence
last light, june 12, 2017
oil on birch panel
8 x 10″

Indigenous water rights & global warming in alberta

david k. laidlaw

Indigenous Nations[1] living in Alberta have existing rights in the waters of Alberta under Canadian law. There may be additional rights, which are discussed below, with an emphasis on treaty rights under Treaty 7 and the Stoney Nakoda First Nations. According to Stoney Nakoda traditions and oral history, the Stoney Nakoda people[2] have always lived on the Great Island and as a branch of the Sioux people they speak the Nakota dialect.[3] Their traditional territories encompassed the foothills and mountains of western Canada from the *Ožadé Tãga* (Brazeau River-Jasper area) down into the *Hũga Baha* (Chief Mountain) area in Montana, west to the *Čã-okiyé Wa-pta* (Columbia River) and east to the Calgary.[4] There were three main bands[5] and before the arrival of Canadians,[6] the Stoney Nakoda people "lived a nomadic way of life, hunting, fishing, and gathering from the abundance of this good land."[7]

The Stoney Nakoda, like other Indigenous Nations, were oriented around an oral culture with a different worldview from Canadians.[8] These differences impacted historical interactions and continue to inform current interactions.[9] Indigenous peoples derived a livelihood from using all aspects of creation[10] with ceremony to express their gratitude to the Creator and to what they harvested, and in that worldview waters were seen as a source of life and cyclical renewal for the resources that were husbanded for future generations.

Indigenous Nations exercised exclusive control over a defined territory with lands and resources *owned* communally by all members and shared in accordance with Indigenous law.[11] Passage or sharing resources in a territory required oral agreement of the controlling Indigenous Nation and a web of diplomatic agreements mediated relations between Indigenous Nations. Those agreements needed periodic renewal to resolve any accumulating differences, particularly after conflicts between Nations, which proliferated during periods of resource scarcity. Alberta had numerous Indigenous Nations prior to contact.[12] How, then, did the government of Alberta come to claim control of the territories and waters of Indigenous Nations?

history

When Canadian settlers arrived in Canada, they found up to two million members of several hundred Indigenous Nations, with varying modes of living, social, and political organizations.[13] Modern-day Alberta was once "owned" by the Hudson's Bay Company [HBC], incorporated by a Royal Charter in 1670[14] that granted monopoly trading rights and ownership for territories bounding the rivers draining into Hudson's Bay[15] — known as Rupert's Land — in return for a nominal rent of "two Elks and two black Beaver."[16] The northern part of Alberta was included in Britain's North-western Territories, described as lands surrounding rivers draining to the Artic.[17]

Canada was organized from the British colonies of Ontario, Quebec, New Brunswick, and Nova Scotia on July 1, 1867 under the *Constitution Act, 1867*,[18] which divided areas of legislative authority between the federal government in section 91 and the provinces in section 92. Canada had jurisdiction over "Indians, and Lands reserved for the Indians" in 91(24) and provincial governments had jurisdiction over property and civil rights in 92(13) and ownership of lands and resources in the provinces in section 109. From 1863 to 1870, Canada negotiated to acquire Rupert's Land and Britain's North-western Territory,[19] resulting in a three-way transaction: HBC surrendered land rights under its Charter to Britain; Britain after accepting the surrender transferred Rupert's Land and its North-western Territory on transfer terms to Canada, who would pay the surrender price to Britain for HBC and gain these lands on July 15, 1870. Aside from the Métis[20] in the Red River Colony who resisted under Louis Riel, Indigenous Nations played no role in this transfer.

Canada, under the transfer terms, negotiated eleven Numbered Treaties with Indigenous Nations, from 1871 to 1921, encompassing most of that territory.[21] In Canadian law, Indigenous Nations were viewed as surrendering their occupancy in territories owned by the Crown.[22] The written text of the Numbered Treaties were based on the Ontario Robinson Treaties from the 1850s, with major terms framed in identical legal language saying that Indigenous Nations surrendered rights to vast tracts of land, in return for promises: that they could continue their traditional way of life on surrendered lands (subject to tracts being taken up by the government); lands would be reserved for their exclusive use (Reserves); as well as annual annuities, supplies, and other benefits.

Treaty negotiations were fraught with misunderstandings.[23] Canadian Treaty Commissioners were representatives of a written culture with legal backgrounds[24] that intended to open lands for settlement as cheaply as possible.[25] Indigenous Nation's leadership were representatives of an oral culture that was faced with dire prospects: threats of unregulated settlement and declining livelihood resources in the evident decline of bison herds. The differing worldviews, legal systems, and languages were a significant barrier to understanding: Treaty Commissioners spoke only English and Indigenous languages had no words or concepts as to the permanent surrender or individual ownership of "land."[26] Indigenous Nations were active, if reluctant, participants in Treaty Negotiations and some of their concerns made it into the written terms of the Treaty.[27]

Indigenous Nations understood the oral promises as terms of the Treaty subject to periodic review.[28] Canadian governments tend to rely on the written text of the Treaty.[29] Growing Canadian settlement led to Treaties being ignored, with the rights of treaty nations not being recognized or "implemented in many, and possibly most, cases." [30]

treaty 7

Treaty 7 was signed on September 22, 1877, between Canada and the Blackfoot Confederacy (Blood, Peigan, and Blackfoot), Tsuu T'ina First Nation (Sarcee), and Stoney Nakoda First Nations encompassing all of Southwestern Alberta in the land surrender provision.[31] According to Stoney Nakoda elders, Treaty 7 was signed primarily as a peace treaty and some mention was made of "sharing 2 feet of the topsoil" but there was no translation as to the legal terms of land surrender.[32] The written terms of Treaty 7 treated the Stoney Nakoda First Nations as one band, with Reserve locations and land amounts always disputed.[33]

Indigenous common law riparian rights

In the prairie-numbered Treaties there was no separate mention of waters aside from Treaty 7 giving the Crown the right on Blackfeet, Blood, and Sarcee Reserves "to navigate, land and receive cargoes on the shores, to build bridges and operate ferries, and use the fords and all the trails leading to" their rivers.[34] This suggests that all other water rights were retained. There was no federal water legislation and the common law doctrine of "riparian rights" applied.[35] There is no property in flowing waters at common law. Riparian rights are part of Canadian common law that set out a number of principles under which the use of surface waters in defined channels (watercourses) are shared by owners of lands next to a watercourse. These principles included:

riparian water rights are automatically vested in owners of land bordering watercourses or through which watercourses flow: those properties are described as riparian property;• riparian property may be obtained by open notorious and continuous occupation for a number of years that gives common law possessory title [to Indigenous Nations];

- water may be used for ordinary domestic purposes connected with the riparian property — regardless of the impact to other downstream riparian property owners;

- water may be used for secondary or "extraordinary purposes" such as irrigation or industrial uses but any waters diverted for these uses must be returned to the watercourse substantially undiminished in quantity and quality, subject to an allowed "reasonable use" diminishment;

- the rights to use water were restricted to the riparian property as they were inseparable from the riparian property.[36]

Riparian principles did not allocate a specific amount of water to riparian property, are not exclusive in nature (as all riparian properties have them) and do not carry a priority — other than being located closer to the water source. These principles, premised on abundant waters, continue to apply in Eastern Canada although they have been overlaid by provincial legislation.

The area covered by Treaty 7 covers the Palliser Triangle, a semi-arid region unsuitable for agriculture.[37] It soon became apparent that agriculture could not thrive in the prevailing conditions and riparian rights to waters were unsuited to the large-scale irrigation necessary for agrarian settlement. In response, Canada passed *The North-West Irrigation Act*, 1894[38] [*NWIA*] attempting to extinguish riparian rights by claiming initially the right to regulate use of all surface waters by Canada by way of a licence.[39] Rights to divert surface water would be licenced originally for three purposes: domestic, irrigation, and other purposes. *NWIA* established a priority system, based on the date of filing. Licences allowed the diversion of water to benefit the land specified and for the purpose(s) listed; however it did not specify a licence duration, and licences were issued in perpetuity in large amounts. Holders of water licences would be entitled to the entire amount of their licence in priority of registration but licences could not deprive the owner of riparian property waters necessary for domestic purposes.[40]

The *NWIA* initially required owners of existing riparian property to obtain a licence for the use of waters but it was amended in 1895 to claim the property in surface waters and to drop the licencing requirement for riparian properties' use of water for domestic purposes (as defined in the legislation).[41] Riparian rights are not rights of the user of water subject to licencing; rather, they are property rights attached to riparian property, and extinguishing property rights requires an express enactment that is absent from the *NWIA* and succeeding federal legislation.[42] The *NWIA* was renamed the *Irrigation Act* and was progressively amended to enlarge the definition of domestic uses to include certain agricultural and industrial purposes.[43]

When Canada created Alberta in *The Alberta Act* (1905), it retained all crown lands and property "in the waters within the province" and instead provided a subsidy for Alberta's government.[44] This differing treatment from the original provinces resulted in a series of identical agreements between Canada and the prairie provinces. *The Alberta Natural Resources Transfer Agreement, 1930* (*NRTA*) transferred Crown lands and the natural resources to Alberta, excepting out federal lands including National Parks and Indian Reserves.[45] The *NRTA* did not, despite mention in the preamble, transfer surface water resources to Alberta.[46] Alberta proceeded to pass *The Water Resources Act*[47] claiming ownership of surface waters in Alberta where "the principles of the *Irrigation Act* were incorporated with only minor drafting amendments" on the basis of provincial ownership only.[48] This error was corrected in *The Natural Resources Transfer (Amendment) Act, 1938* with a *Memorandum of Amendment* transferring "the interest of the Crown in the waters ... under the *North-West Irrigation Act*, 1898" backdated to 1930.[49]

Riparian rights are property rights, and any provincial legislation after 1930 extinguishing or regulating them on Reserves is outside of the province's legislative authority and invalid.[50] Reserve land straddling or adjoining watercourses are riparian property and Indigenous Nations have riparian rights to divert surface water in unlimited amounts *without* requiring any licence for domestic purposes as defined in the *Irrigation Act* — including common law riparian rights and legislated reasonable consumptive use for agricultural machinery and industrial purposes. The Stoney Nakoda First Nations' Reserves straddle many waterways, including the Mini Thni (Bow River) flowing through Calgary.

treaty and Aboriginal rights to water?

Treaties are governed by the *honour of the crown* requiring consideration and fidelity to Treaty Indigenous Nations with special interpretative and implementing principles for Treaties and legislation affecting Treaty rights. Since April 17, 1982, Treaties are protected in the *Constitution Act, 1982* by section 35(1) where "the *aboriginal* and *treaty* rights of the aboriginal peoples of Canada are hereby recognized and affirmed."[51] Aboriginal rights are activities central to the lifestyle of Indigenous Nations, being practised in a current form that relates to the original practice (prior to Canadian contact), which have not been extinguished by explicit legislation prior to April 17, 1982.[52] Land-based practices that qualify as aboriginal rights can form site-specific aboriginal rights on Crown lands subject to the applicable treaty.[53] Aboriginal and Treaty rights cannot be restricted unless there is a valid legislative object, such as public safety or conservation, and the restriction must accord with historical relationship between Canada and aboriginal peoples including: the honour of the crown, and where the government has assumed control over a central aboriginal interest such as land, the fiduciary duty of the Crown to uphold aboriginal interests.[54] Aboriginal rights take priority over other uses as they pre-date those uses, thus in circumstances of a constrained resource — such as water — aboriginal rights to use water take priority over all other uses.[55]

Treaties can embody some aboriginal rights and provide additional rights.[56] They are not a complete code, as aboriginal and treaty rights can be practised in the same territory unless barred by the Treaty. Treaties can, in Canadian jurisprudence, surrender aboriginal title and rights in the territorial surrender language, if any, in accordance with the Treaty terms with Reserves remaining aboriginal title lands.[57] The exercise of Treaty and aboriginal rights are not limited to traditional practices; they may be exercised in a modern form together with ancillary rights, if there is a continuity between the date of the Treaty or pre-contact practices. Ancillary rights include those rights necessarily incidental to the exercise of the protected rights and rights reasonably incidental to the protected rights.[58]

The numbered Treaties in the prairie provinces[59] were intended to promote the transition of Indigenous Nations to a settler-agrarian lifestyle of farmers and ranchers for assimilation into Canadian society.[60] Indigenous pre-contact agricultural practices such as horse breeding and care were common on the prairies, and pre-dated Treaty signing. Treaty 7 offered agricultural supplies "for the encouragement of the practice of agriculture among the Indians."[61] Ranching and farming require water supply security and may form the basis for a Treaty right to water.

The prairie numbered Treaties[62] were signed prior to the passage of the *NWIA*'s amendments in 1895 that arguably appropriated the property in surface waters. This contemporaneous process of signing Treaties and the passage of the *NWIA* has led scholars to question the *NWIA* affecting Indigenous rights to water, as that would "entail a highly disenchanted view of federal policy to conclude, that the federal government, in the midst of treaty negotiation, engaged in so substantial a violation of the treaty promises."[63]

The numbered Treaties, encompassing all of Alberta's territory, preserved the right of Indigenous Nations to obtain a traditional livelihood in the surrendered territories on Crown lands not "taken up" and assigned to an incompatible use. In Clause 12 of the *NRTA,* Canada required Alberta to fulfill Treaty obligations on the lands and resources transferred to it. The Courts have interpreted the *NRTA* as a constitutional document, modifying the Treaties to expand Treaty rights over the entire province but eliminating the exercise of Treaty rights for commercial purposes.[64] Alberta's position, expressed in its "Policy on Consultation with First Nations,"[65] is that Treaty rights are limited to the "rights to hunt, fish and trap for food."[66]

Treaty rights, even on the limited basis acknowledged by Alberta, carry within them a number of necessary or reasonably incidental Treaty rights. These include adequate water supply for transportation and access to hunting, fishing, and trapping areas, as well as adequate water quality to support the supply of wildlife, fish, and fur-bearing animals.[67] Treaty rights to water supply and quality have not been adjudicated in Alberta, although there are a number of lawsuits including them.[68] These Treaty rights would apply off-Reserve, and may affect provincial laws through the application of section 88 of the *Indian Act*, which limits provincial laws under a Treaty.

water in alberta

Canada is often seen as a water rich country with 20 per cent of the world's fresh water. However, Canada holds only 6.5 per cent of the global supply that is renewable; the balance is stored in lakes, underground aquifers, and glaciers. Of that renewable water supply, 60 per cent drains northward into the Arctic Ocean and Hudson Bay. As a result, it is unavailable to the 85 per cent of the Canadian population who live along the country's southern border. Alberta holds approximately 2.2 per cent of Canada's freshwater, but 80 per cent of Alberta's freshwater supply is found in the northern reaches of the province while 80 per cent of water demand lies in the south. Only 13.3 per cent of Alberta rivers drain south and east into Hudson's Bay with 86.6 per cent draining north to the Artic Ocean.[69]

Alberta's water system, inherited from the *NWIA/Irrigation Act*, involves: prior allocation by government licences giving landowners the right to use large amounts of water annually on their land, subject to limited domestic use by non-Indigenous riparian land owners. The priority is based on the time of registration, such that in times of water shortage, the earliest registration can use its entire allocation before a subsequently registered water licence receives any. Historically governments have allocated the majority of licences for agricultural uses and this remains the largest use today.[70] In 2009 Alberta licenced a total of 9.89 billion m^3 water, of which 97 per cent (9.59 billion m^3) is from surface water sources and only 3 per cent (301 million m^3) of the volume is from groundwater sources. Water is allocated to 187,551 licence holders: 49,376 (~26 per cent) are groundwater licence holders, and 138,175 (~74 per cent) are surface water licence holders. The current allocation total as of 2016 is 10.198 billion m^3 with essentially the same distribution of uses.[71] Groundwater originates from precipitation soaking down into the ground, and while groundwater reserves may dwarf surface waters only 1 per cent of those reserves are in usable aquifers. Use of groundwater will not address water shortages given the limited recharge rate of aquifers.

Not all uses are equal. Agricultural uses are consumptive, as most of the water is incorporated into the products of agriculture such as grains or livestock. Other uses are less consumptive, as municipalities will return 80–90 per cent of the water; the commercial water used for cooling is entirely returned and other uses do not incorporate water to the same extent as agriculture. All consumptive uses generate pollution, affecting water quality for downstream users and environments; removing that pollution requires sewage and water treatment facilities. Environmental uses are supportive, as they represent water allocation retained in the sources as instream flows to maintain aquatic and riverine bank ecosystems.

The most extensive drought in Alberta history, the dustbowl era of the 1930s, did not lead to changes in the allocation system. Instead, Alberta embarked on an effort to increase the useable water supply by storing water behind dams. It is estimated that $1B was devoted to water projects from 1930–90, with Alberta continuing to issue extensive water licences.[72] Suitable storage sites for major water projects, at least in the south of Alberta, have been exhausted.

The current *Water Act* (1995) has modified the allocation system, by among other things: requiring the development of basin-wide water management plans, the power to close water basins, new water licences issued on a conditional basis and limited to renewable five-year terms, and unlinking water licences from land to encourage a market for water licence transfers intended to drive conservation by indirect water pricing. Any approved water licence transfer leads to two new conditional water licences with the same priority number as the original but a potential reduction in water amount of up to 10 per cent for in-stream flows. This transfer mechanism has discouraged senior licence holders, holding the bulk of water licences, from participating in water licence markets. These *Water Act* measures can provide a "safety valve" in transferring uses — but are not likely to increase the water supply from conservation.

Alberta is divided into seven watershed basins named after the major river they drain into: the Hay River, Peace/Slave, Athabasca, Beaver, Milk, and the North and South Saskatchewan River Basins, with multiple water management plans. Each basin has varying distribution of licenced uses and water consumption; for example, in 2009 Alberta allocated 2.74 billion m^3 of water from the Bow River, with 73 per cent used for irrigation and 17 per cent used by the City of Calgary, and from the North Saskatchewan River 2.01 billion m^3 of water, with 79 per cent used in thermal power plants and 7 per cent used in the city of Edmonton.[73]

Due to concerns about over-allocation and insufficient in-stream flows, Alberta stopped issuing new licences in the Oldman, Bow, and South Saskatchewan sub-basins in August 2006 in the *Approved Water Management Plan for the South Saskatchewan River Basin (Alberta)* (2006) [SSRB Water Management Plan].[74] Alberta made some exceptions for new in-flow stream allocations, water storage releases, and First Nation licences.[75] This closure has led to growing conflict between water users and it is estimated that, even with water conservation measures reducing water consumption by 30 per cent,[76] municipalities in the Calgary region will exhaust their own and Calgary's current extensive water allocation by 2030.[77]

Land use is inextricably tied to waters and the use of land has been regulated in Alberta since the early 1900s.[78] The current legislation, the *Alberta Land Stewardship Act*[79]

[*ALSA*] was passed in 2009 and is derived from the *Land Use Framework* (2008)[80] that divided Alberta's territory into seven Regions corresponding generally with drainage basins of Alberta's major rivers.[81] *ALSA* provides a mechanism to plan for the direction of desired economic, environmental, and social objectives by way of regional plans that are expressions of government policy approved by Cabinet. Provincial municipalities and decision-making bodies are required to amend their bylaws and policies to ensure compliance.[82] The government-controlled public consultation, protection of regional plans by Cabinet secrecy, and extensive government discretion have been criticized by environmental groups, First Nations, and in academia — particularly with the cumulative impacts of development.[83]

The South Saskatchewan Regional Plan [SSRP][84] is the second regional plan approved by Cabinet effective September 1, 2014.[85] The SSRP encompasses the area described in 2007[86] as "water short," namely the South Saskatchewan River Basin [SSRB]. The SSRP noted that the pressure on water resources is significant with over 20,000 water licences. The major use for these licences is agriculture, which accounts for 75 per cent of total water allocation combined with a growing population, currently at 1.8 million people, and while current actual use is 55 per cent for municipalities and 66 per cent for agricultural licences, those actual uses will grow. This is particularly compounded by periodic phases of natural low flow and drought in the region.[87]

The SSRP affirmed the SSRB Water Management Plan as a provincial strategy that recognizes the limit of water resources has been reached in those watersheds, thus any decision-making bodies must review their rules and procedures to ensure compliance with the SSRP. SSRP says, in an understatement, that the challenge of matching water supply to demand will be compounded by changing climactic conditions.[88]

global warming

Global warming is real, it is caused by humanity, and it represents a significant threat to humanity and the environment in the near future. All nations have agreed in the *Paris Accords* (2015)[89] that climate change is an urgent threat and a common concern of humankind, and have undertaken to hold "the increase in the global average temperature to well below 2°C above pre-industrial levels," and likewise pursue "efforts to limit the temperature increase to 1.5°C above pre-industrial levels." This would significantly reduce the risks and impacts of climate change.[90]

Globally the last three decades have been successively warmer at the Earth's surface than any preceding decade since 1850. The period from 1983 to 2012 was the warmest thirty-year period in the last 1400 years in the Northern Hemisphere. The globally averaged combined land and ocean surface temperature data as calculated by a linear trend show a warming of 0.85°C over the period 1880 to 2012.[91] The Northern Hemisphere warms faster than the global average because it has more land and less ocean water than the Southern Hemisphere (water warms slowly).

In Alberta, over the past 100 years the mean temperature has increased by 1.4°C with most of the increase occurring since 1970. Between 1912 and 2011, the average annual temperature increased by 1.1°C (0.1 per decade) in the southern half of the province and double that (2.3°C or 0.2 per decade) in the north. Since 1970 the pace of warming has intensified increasing at a rate of 0.3°C per decade in both the north and the south.[92]

River summer flows are important for irrigation in Alberta, as demand is higher in summer. In 2000 these flows have dropped in the South Saskatchewan River at Medicine Hat to 53.8 per cent, North Saskatchewan River at Prince Albert to 66.6 per cent, Peace River at Peace River to 62.3 per cent, and Oldman River at Lethbridge to 59.3 per cent from the 1910 summer flows. The summer flow reduction further downstream is even more severe, with measurements at Saskatoon being 20 per cent of the 1910 flows.[93] Some, but not all, of this decline in summer flows can be attributed to the extensive dam projects built between 1930 and 1990, with 50 per cent of flows being moderated by those water projects but only 25–50 per cent of the average river discharge being accounted for.[94]

Cumulative emissions of CO_2 largely determine global mean surface warming into the late twenty-first century and beyond; reductions in global cumulative GHGs will take time, as GHG's do not dissipate rapidly. Even with reductions in current emissions, significant changes in trends will not occur until 2050.[95] This means temperature increases in Alberta from the temperature ranges in 2000, range from a projected high of >2.71 °C to a minimum 2.19 °C, with consequent moisture loss of 2.4 to 18 per cent. Southern Alberta has the higher ranges of temperature increases and moisture loss.[96]

One of the most visible impacts of climate change in Alberta is glacial retreat, most evident in the accelerating retreat of the Athabasca Glacier in the Columbia Icefield between Jasper and Banff which is losing ~16,000,000 cubic metres of ice each year. Studies estimate that glacial melt contributes an average of 0.6 per cent of the annual flow in the SSRB and about 2.4 per cent of the base flow in the Bow River at Calgary, although in low-flow years, the percentages could be significantly higher.[97] The SSRB lost half of the glacier area between 1975 and 1998; this loss of glaciers will lead to increased springtime flows and lower summer flows as glaciers, particularly in the Bow River basin, moderate river flows by retaining the winter snow and feeding meltwater into rivers into the crucial summer flows.[98]

There have been significant advances in climate change modelling on a global and regional basis since 2009.[99] In Alberta, some recent representational studies have shown:

> increase in winter and early spring flows, declines in summer (about 15 per cent) and annual flows (about 5 per cent) in the Oldman River Basin, from Shepherd et al., 2010.[100]
>
> Increases in both high and low flow magnitudes and frequencies, large increases to winter and spring streamflow are predicted for all climate scenarios. Spring runoff and peak streamflow occur up to four weeks earlier than in the 1961–90 baseline for the Cline River Basin (North Saskatchewan) (Kienzlea et al., 2012);[101] and
>
> In the 2050s and 2080s, southern Alberta will be expected to experience more frequent and severe intensive storm events in the May, June, July, and August season that could potentially increase the risk of future flooding in this region. (Gizaw et al., 2016).[102]

Alberta will see significant economic and environmental impacts, between now and 2050 including:

> infrastructure costs to protect towns and cities from flooding with higher spring flows and upgrading irrigation facilities to accommodate lower summer flows;
>
> environmental impacts including: increased flow variability affecting riverine ecosystems (the most threatened ecosystem in Alberta); warmer waters threatening marine life and temperature-driven biome movement northward;
>
> drought due to greater soil evaporation, lower recharge rate of rivers, leading to potential desertification in southern areas affecting agriculture incomes; and

> water supply conflict with lower flows in summer leading to less "useable water" for irrigation and needs from an expanding population.

If we do nothing to lower CO_2 releases, Alberta in 2100 will see a doubling in CO_2 concentrations predicted to cause a 6–8°C warming and a resulting decrease in soil moisture, predicted to be between 20 and 40 per cent with "challenging" consequences for Albertans.[103]

There have been proposals for additional water supply within the current regulatory systems, but Amec's Report, "Water Storage Opportunities in The South Saskatchewan River Basin in Alberta (2014)," concluded there was limited opportunity to do so.[104] The original study leading to entrenchment of the SSRB Water Management Plan provided at best a thirty-year window before water shortages developed, but the proposed water licence market, intended to drive water conservation, has been frustrated by the historical distribution of water allocations.[105]

Change to the Alberta's water allocation system is needed. A proposal by Professor Arlene Kwasniak entitled *Climate Change and Water: Law and Policy Options for Alberta* (2017)[106] outlines transition to a new framework that would respect the environment, encourage sustainable development, and be climate change resilient. The development of these changes requires the participation of Indigenous Nations, not only to clarify and affirm their pre-existing rights to water, but also to access the stewardship worldview that has allowed them to live in a sustainable manner in Alberta for thousands of years.

notes

Websites are current to February 1, 2018. Case law and legislation may be foundat The Canadian Legal Information Institute (CanLII)'s website at: **canlii.org**

1. Indigenous People living in Canada prefer the name for themselves in their language and are mostly indifferent to the Canadian name accorded to them, although they may describe themselves as a First Nation to emphasize their political status and priority. The modern collective term is Indigenous.

2. In Nakoda they are known as the *Îyãhé Nakoda*, previously as the *Iyethkabi*. At Rocky Mountain Nakoda website: <http://www.rockymountainnakoda.com>. They are also known as *Yéθka*, *Yéθkabi*, or *Iyéθka*, which means "*Those Without Blemish*," and historically they were known as the *Wapamakθé*, meaning "Head Decapitators."

3. *These Mountains Are Our Sacred Places*, Chief John Snow (Toronto: Samuel Stevens, 1977) [Snow, *Mountains Are Our Sacred Places*], 2. Great Island is an English translation of their Indigenous name for North America.

4. Ibid. According to Rocky Mountain Nakoda website, supra n2, their territory goes north to *Osoda Wa-pta* (Smoky River), south to *Îdukabi Tĩda* (tobacco plains) in Montana, and east to *Wotawa Baha* (Cypress Hills).

5. Ibid., 4. Now called the Goodstoney/Wesley in the north, the Bearspaw in the south, and the Chiniki in the middle.

6. The use of "Canadians" is deliberate. Current Canadian residents have, since 1867, inherited the territories, re sources, *and* obligations of Britain arising from historical encounters with Indigenous Peoples, as well as incurring new obligations. Current Canadian residents may not have participated in the history of Indigenous Peoples' suppression and dispossession but they live in a Canadian society *that has prospered on that history*.

7. Snow, *These Mountains Are Our Sacred Places*, supra n3, 2.

8. This difference may be attributable to the respective creation stories: in both versions the Creator made the world and everything in it, but in Canadian Biblical stories the Creator (God) expelled humans from his presence in Paradise and gave humans dominion over plants and every living thing on Earth whereas in Indigenous creation stories the Creator remained in the world with his creations and there was no distinction between them. Snow, *These Mountains Are Our Sacred Places*, supra n3, 1–13, describes the Stoney Nakoda creation stories and the consequent social organizations of the Stoney Nakoda Nations. See generally: Sandra Tomsons and Lorraine Mayer, eds., *Philosophy and Aboriginal Rights: Critical Dialogues* (Don Mills: Oxford University Press, 2013) and Rupert Ross, *Dancing with a Ghost: Exploring Indian Reality* (Markham: Reed Publishing, 1992).

9. *Report of the Aboriginal Justice Inquiry of Manitoba* (1999) at <http://www.ajic.mb.ca/volume.html> [*Manitoba Aboriginal Justice*] volume 1, chapter 2 at <http://www.ajic.mb.ca/volumel/chapter2.html>. A basic summary, *Differences between Traditional Aboriginal Cultures and Mainstream Western Culture*, can be found at <http://www.med.uottawa.ca/SIM/data/Images/Aboriginal_x_Western_values.pdf>.

10. Contrary to stereotypes, Indigenous Nations traded portable, high-value resources including *underground re sources*, such as chalk, flint, copper, and hydrocarbons (pitch) gathered from sites where their technology made them accessible through "kin trade" networks spanning North America. See Kerry Abel, ed., *Aboriginal Resource Use in Canada* (Winnipeg: University of Manitoba Press, 1991).

11. Sidney L. Harring, *White Man's Law: Native People in Nineteenth-Century Canadian Jurisprudence* (Toronto: University of Toronto Press, 1998) [Harring, *White Man's Law*], 129–31. See also Edwin Thompson Denig, *The Assiniboine*, ed. J.N.B. Hewitt, (Regina: Canadian Plains Research Center, 2000), 82–4.

12. A map of pre-contact Indigenous language groups is in volume 1 of the *Report of the Royal Commission on Aboriginal People* (Ottawa: Supply and Services Canada, 1996) [*Aboriginal Peoples Report*], 21. Online: <https://qspace.library.queensu.ca/handle/1974/6874>. A colourized version can be found at

<http://www.amauta.info/satellite/cai.jpg>. An interesting but incomplete description by the Alberta Government, *Aboriginal Peoples of Alberta: Yesterday, Today, and Tomorrow* is online at <http://indigenous.alberta.ca/documents/AboriginalPeoples.pdf?0.051602806285153124>.

13. *Aboriginal Peoples Report*, ibid., contains an extensive history. See *Manitoba Aboriginal Justice*, supra n9, volume 1, chapter 3 at <http://www.ajic.mb.ca/volumel/chapter3.html> for a historical overview.

14. Hudson's Bay Company History Foundation, online at <http://www.hbcheritage.ca/things/artifacts/the-charter-and-text>.

15. Ibid., 1, para 3. This ownership was justified on the basis of *first discovery*, a doctrine that originally excluded claims from other European powers which was transformed to say that the discovering country could, by settlement if the area was unoccupied (*terra nullius*), obtain title, or if occupied by conquest or agreement with the original occupants (*cession*). Canada did not engage in Indian Wars and no lands were acquired by conquest. *Tsilhqot'in Nation v British Columbia*, [2014] 2 SCR 257, 2014 SCC 44 (CanLII), at para 69 notes that the doctrine of *terra nullius* did not apply in Canada because of the *Royal Proclamation, 1763* (an accurate transcription is in *Aboriginal Peoples Report* volume 1 appendix D) which came after the British and Indigenous Nation allies conquered Québec. After the surrender of French colonies, the British Government issued the *Royal Proclamation, 1763*, which preserved lands beyond the "frontier" as "Hunting Grounds" for the Indigenous Nations *and* any unsurrendered lands within the colonies. It established a Crown monopoly on land purchases (cession) from Indigenous Nations saying purchases must be approved at a public meeting of the "Indians" held for that purpose.

16. Ibid., 3, para 8. This Rent was to be paid whenever the Monarch came to that territory and has been performed four times: <http://www.hbcheritage.ca/history/fur-trade/the-rent-ceremony>.

17. *An Act to make further Provision for the regulation of the Trade with the* Indians, *and for the Administration of Justice in the North-western territories of* America (1859) 22 & 23 Vic, c 26 (UK). Online: <http://eco.canadiana.ca/view/oocihm.9_01245/2?r=0&s=1>.

18. *The Constitution Act, 1867*, 30 & 31 Vict, c 3. (*Constitution Act, 1867*).

19. A map is in Frank J. Tough, "Aboriginal Rights Versus the Deed of Surrender: The Legal Rights of Native Peoples and Canada's Acquisition of the Hudson's Bay Company Territory" *Prairie Forum* 17, no. 2 (1992): 225-6. This is the largest land purchase in the world and contains 75 per cent of Canada's territory.

20. The Métis people are descendants of fur traders and Indigenous women, who became a separate people acting as intermediaries in the fur trade. See Sylvia Van Kirk, *Many Tender Ties: Women in Fur-Trade Society, 1670-1870* (Winnipeg: Watson & Dwywer, 1999). They are now recognized as one of Canada's aboriginal peoples.

21. Historical Indian Treaties of Canada, Atlas of Canada Reference Maps online: <http://ftp.geogratis.gc.ca/pub/nrcan_rncan/raster/atlas_6_ed/reference/eng/treaties.jpg>.

22. The Privy Council, Canada's highest court until 1949, decision in *St. Catherines Milling and Lumber Co v The Queen* (1888), 14 App Cas 46 (PC), described "Indian title" as a "personal and *usufructuary* right [to harvest the land owned by another], dependent upon the good will of the Sovereign [under the *Royal Proclamation, 1763*] ... that there has been all along vested in the Crown a substantial and paramount estate, underlying the Indian title, which became a *plenum dominium* [complete ownership] whenever that title was surrendered or otherwise extinguished," 54-5. Harring, *White Man's Law*, supra n11, describes this case, 135-47. This did not mean the sur render by Treaty comported with Indigenous law, which "generally did not have a concept of land ownership that would have included authority to transfer absolute title to the Crown." In Kent McNeil, "Extinguishment of Aboriginal Title in Canada: Treaties, Legislation, and Judicial Discretion," *Ottawa Law Review* 301, no. 33 (2001-02): 304-8. Douglas Harris, *Fish, Law, and Colonialism: The Legal Capture of Salmon in British Columbia* (Toronto: University of Toronto Press, 2001), 18-27 and 62-5, argued that dividing lands/"hunting grounds"/livelihood re sources was contrary to Indigenous law as they could be *shared* with various peoples, including Canadians, but not divided.

23. Canadian contemporary views: Alexander Morris, *The Treaties of Canada with the Indians of Manitoba and the North-West Territories, Including the Negotiations on Which They Were Based, and Other Information Relating Thereto* (Toronto: Belfords, Clarke & Co., 1880). Online: <http://www.gutenberg.org/ebooks/7126>; Indigenous views: Richard T. Price, ed., *The Spirit of the Alberta Indian Treaties*, 3rd ed. (Edmonton: University of Alberta Press, 1999) [Price, *Spirit of Alberta Indian Treaties*]. An American historian's view: Harring, *White Man's Law*, supra n11, 250–70, with a description of misunderstandings in Treaty Negotiations.

24. Alexander Morris was the main Treaty Commissioner; a lawyer, judge, and politician, he was also actively involved in treaty negotiations with aboriginal groups, signing Treaties 3, 4, 5, and 6, and revising Treaties 1 and 2.

25. Reserves in the Prairie region were limited to one square mile for each family of five persons, or in that proportion with minimal annual annuities. There was some flexibility in the amount of supplies and annual gifts.

26. *Aboriginal Peoples Report*, supra n13, volume 2, chapter 5: "Many of the [historical] treaties ... were made with one of the parties (the Crown) believing that the central feature of the treaty was the purchase or extinguishment of the other party's Aboriginal title, while the very idea of selling or extinguishing their land rights was beyond the contemplation of the Aboriginal party, because of the nature of their relationship to the land." See John Leonard Taylor, "Two Views on the Meaning of Treaties Six and Seven," 9–46, and Richard Daniel, "The Spirit and Intent of Treaty Eight," in Price, *Spirit of Alberta Indian Treaties*, supra n23, 47–102.

27. Notably in Treaty 6's mention of a "medicine box," but also education promises for Reserve schools.

28. They have passed that oral knowledge to their descendants. See Bruce G. Miller, *Oral History On Trial: Recognizing Aboriginal Narratives In The Courts* (Vancouver: University of British Columbia Press, 2011).

29. *Smith v The Queen*, [1983] 1 SCR 554, 1983 CanLII 134 (SCC), 568–70.

30. *Aboriginal Peoples Report*, supra n13, vol. 2, 48.

31. A transliteration of Treaty 7 is at <http://www.aadnc-aandc.gc.ca/eng/1100100028793/1100100028803>. The Federal Government had commissioned research Historical Treaties during the late 1980s. Online: <http://www.aadnc-aandc.gc.ca/eng/1100100028653/1100100028654>

32. Walter Hildebrandt, *The True Spirit and Original Intent of Treaty 7* (Montreal: McGill-Queen's University Press, 1996), 77–80. The North-West Mounted Police had ended American fur traders' provision of alcohol to Indigenous people, and the presence of their commander at the treaty negotiation was influential.

33. A 2012 Amended Specific Claim for a Treaty Land Entitlement argues that they are entitled to an additional 25,472 acres. Online: <http://www.scttrp.ca/apption/cms/UploadedDocuments/20126001/018-SCT-6001-12-Doc12.pdf>. Historically Reserve surveys excluded productive lands, especially mineral resources. See Claudia Notzke, *Aboriginal Peoples and Natural Resources in Canada* (Lethbridge: Captus Press, 1994), 203–4.

34. This is from the first provision setting aside Reserves of Treaty 7. See supra n31.

35. J.E. Cote, "The Reception of English Law," *Alberta Law Review* 15 (1977): 29.

36. Alistair R. Lucas, *Security of Title in Canadian Waters* (Calgary: Canadian Institute of Resources Law, 1990), 5–8. The rules for acquisition and use of underground waters are different (8–11).

37. The Palliser Triangle named for Captain John Palliser, the leader of 1857–59 Official Survey Expedition to Canada's west. A map can be found online at <https://commons.wikimedia.org/wiki/File:Palliser%27s_Triangle_map.png>.

38. *The North-West Irrigation Act*, 1894, 57–8 Vic c 30 [*NWIA*].

39. Ibid,. s 4. Richard H. Bartlett, *Aboriginal Water Rights in Canada: A Study of Aboriginal Title to Water and Indian Water Rights* (Calgary: Canadian Institute of Resources Law, 1990) [Bartlett, *Aboriginal Water Rights*], 152.

40. Ibid., s 9.

41. *An Act to amend the North-West Irrigation Act,* (1895), 58–9 Vic c 33, [*NWIA* Amendments, 1895]. It continued the licencing requirement for riparian owners' non-domestic uses. David R. Percy, "Water Rights in Alberta," *Alberta Law Review* 15 (1977) [Percy, Water Rights in Alberta]: 156–7.

42. Monique M. Passelac-Ross and Christina M. Smith, *Defining Aboriginal Rights to Water in Alberta: Do They Still "Exist"? How Extensive Are They?* (Calgary: Canadian Institute of Resources Law, 2010) [Ross and Smith, *Abo riginal Rights to Water in Alberta*], 35–6; Percy, "Water Rights in Alberta," supra n41, 150–60; Bartlett, *Aboriginal Water Rights*, supra n39, 155–63.

43. *Irrigation Act*, RSC 1906 c 61. and subsequently *Irrigation Act, 1927*, RSC 1927. [*Irrigation Act*]. This would include household, sanitary, stock watering, and operation of agricultural and industrial machinery. The *Irrigation Act* remained in force until 1952 but did not apply in the Western provinces after the 1930s. Percy, "Water Rights in Alberta," supra n41, 146.

44. *The Alberta Act*, SC 1905, c 3, s 21. This was to maintain Canada's control over prairie settlement.

45. The NRTA was implemented by legislation, *An Act respecting the transfer of the Natural Resources of Alberta*, SA 1930, c 21, and *An Act respecting the transfer of the Natural Resources of Alberta*, SC 1930, c 3.

46. Ibid. The *NRTA* Clause 8 entitled *Water* dealt with the ownership of hydropower dams.

47. *The Water Resources Act*, SA 1931, c 71.

48. Percy, "Water Rights in Alberta," supra n41, 146 and 158.

49. *The Natural Resources Transfer (Amendment) Act, 1938*, SC 1938, c 36, and *An Act to Ratify a certain Agreement between the Government of the Dominion of Canada and the Government of the Province of Alberta*, SA 1938, c 14. [*NRTA Amendment, 1938*]. Underground water resources were not transferred to Alberta. Ross and Smith, *Aboriginal Rights to Water in Alberta*, note among other things that the *NRTA* did "not abrogate the Indian interest in reserve lands nor the federal government's right to administer such lands," supra n42, 36–40.

50. *Re Stony Plain Indian Reserve No. 135*, [1982] 1 WWR 302, 314–15. Provincial legislation of this nature is *ultra vires* – that is shorthand for being outside of the provincial legislative powers in section 92.

51. *R v Sparrow,* [1990] 1 SCR 1075, 1990 CanLII 104 (SCC) interpreted this language.

52. *R v Van der Peet*, [1996] 2 SCR 507, 137 DLR (4th) 289, *R v Sappier; R v Gray,* 2006 SCC 54, [2006] 2 SCR 686 In *R v Powley,* 2003 SCC 43, [2003] 2 SCR 207, Métis rights originated in the contact to *pre-control* period when Canadians achieved political and legal control in an area. *Sparrow*, supra n51 set the requirements to extinguish aboriginal rights by explicit legislation prior to 1982.

53. *R v Adams*, [1996] 3 SCR 101, 138 DLR. (4th) 657 [*Adams*].

54. *Sparrow*, supra n58, *Wewaykum Indian Band v Canada,* [2002] 4 SCR 245, 2002 SCC 79 (CanLII), *Manitoba Métis Federation Inc v Canada (Attorney General)*, [2013] 1 SCR 623, 2013 SCC 14 (CanLII).

55. *Sparrow*, supra n51, 115–16.

56. *Adams*, supra n53, 49. E.g. Treaty rights to annuities, supplies, and healthcare in Treaty 6's "medicine box."

57. *Guerin v The Queen*, [1984] 2 SCR 335, 13 DLR (4th) 321. Subject to regulation under the *Indian Act.*

58. Ross and Smith, *Aboriginal Rights to Water in Alberta*, supra n42, 22.

59. Historical Treaty Map supra n21. Treaty 8 excepted as lands area were inhospitable to agriculture.

60. Bartlett, *Aboriginal Water Rights*, supra n39, 20; see generally Sarah Carter, *Lost Harvest: Prairie Indian Reserve Farmers and Government Policy* (Montreal and Kingston: McGill-Queen's University Press, 1990).

61. Stoney Nakoda First Nations were one of the few Treaty 7 Indigenous Nations to choose agrarian supplies.

62. With smaller areas in subsequent adhesions to Treaty 5 in 1908 and Treaty 6 in 1899.

63. Bartlett, *Aboriginal Water Rights*, supra n39, 154–64. This conclusion is bolstered by the *honour of the crown* as a constitutional principle in interpreting Treaties.

64. *R v Badger*, [1996] 1 SCR 771, *Frank v The Queen*, [1978] 1 SCR 95, and *R v Horseman*, [1990] 1 SCR 901.

65. *The Government of Alberta's Policy on Consultation with First Nations on Land and Natural Resource Management, 2013*. Online at <http://indigenous.alberta.ca/documents/GoAPolicy-FNConsultation-2013.pdf?0.45142952131362>.

66. Ibid., appendix 2, 61. It considers any other practices as "traditional uses" being "customs and practices on the land that *are not existing section 35 Treaty rights*. We have argued that is a fundamental misinterpretation of the Treaties in David K. Laidlaw and Monique Passelac-Ross, *Alberta First Nations Consultation & Accommodation Handbook*, CIRL Occasional Paper #44 (Canadian Institute of Resources Law, Calgary, 2014) online at <http://cirl.ca/system/files/ConsultationHandbookOP44w.pdf>.

67. Ross and Smith, *Aboriginal Rights to Water in Alberta*, supra n42, 28.

68. Ibid., 28–30, has a comparison of claims in other jurisdictions and suggest those decision would apply in Alberta.

69. Alberta Environment Information Centre, *Facts About Water in Alberta* (Edmonton: Government of Alberta, 2010) [*Facts About Water in Alberta*], 5–6. A map is on page 12, at <http://aep.alberta.ca/water/water-conversation/documents/FactsAboutWaterAlberta-Dec2010A.pdf>.

70. Ibid., 35.

71. Ibid., 33–5. Information about the 2016 licenced water allocation by industry is available at the Alberta Energy Regulator [AER] website: <https://www.aer.ca/data-and-publications/water-use-performance#report>.

72. David R. Percy, "Seventy-Five Years of Alberta Water Law: Maturity, Demise & (and) Rebirth," *Alberta Law Review* 35 (1996): 227.

73. Facts about Water, supra n69, 37.

74. *Approved Water Management Plan for the South Saskatchewan River Basin (Alberta)* (2006) online <http://aep.alberta.ca/water/programs-and-services/river-management-frameworks/south-saskatchewan-river-basin-approved-water-management-plan/documents/SSRB-ApprovedWaterManagementPlan-2006.pdf>.

75. *Bow, Oldman and South Saskatchewan River Basin Water Allocation Order*, Alta Reg 171/2007 authorizes the granting of junior water licences to certain Indigenous Nations. Alberta is attempting to compel them to obtain a provincial water licence by refusing project approvals. See Matthew McClearn, "Showdown over Water Looms for Alberta Reserves," *Globe & Mail*, October 16, 2016.

76. A target set on the basis that this goal would not substantially affect modern lifestyles in the *Water for Life*

Policy: Alberta's Strategy for Sustainability (2003) online at <http://aep.alberta.ca/water/programs-and-services/water-for-life/strategy/documents/WaterForLife-Strategy-Nov2003.pdf>.

77. David J. Pernitsky and Natalie D. Guy, "Closing the South Saskatchewan River Basin to New Water Licences: Effects on Municipal Water Supplies," *Canadian Water Resources Journal* 35, no. 1 (2010): 90.

78. Bernard J. Roth and Rachel A. Howie, "Land-Use Planning and Natural Resource Rights: The Alberta Land Stewardship Act," *Journal Energy & Natural Resources Law* 29 (2011): 473–6. History of Planning Legislation, online at <http://propertyrightsguide.ca/history-of-planning-legislation-in-alberta/>. Sara L Jaremko, *A Critical Exploration of the South Saskatchewan Regional Plan in Alberta* (Calgary: Canadian Institute of Re sources Law, 2016), 7–13 [Jaremko, *Critical SSRP*]. Online: <https://dspace.ucalgary.ca/retrieve/46560/South SaskPlanOP54w.pdf>.

79. *Alberta Land Stewardship Act*, SA 2009, c A-26.8 [*ALSA*] The Alberta Environment and Parks maintains a website at <https://landuse.alberta.ca/ResultsResources/Pages/MapsandShapefiles.aspx>.

80. *Land Use Framework* (2008). Online: <https://landuse.alberta.ca/LandUse%20Documents/Land-use%20Frame work%20-%202008-12.pdf>. Alan Harvie and Trent Mercier, "The *Alberta Land Stewardship Act* and its Impact on Alberta's Oil and Gas Industry," *Alberta Law Review* 48 (2010–11): 295.

81. *ALSA* was hailed as unique in the common law world and promoted as a way to address cumulative effects of development in Roth and Howie, "Land-Use Planning and Natural Resource Rights," 472.

82. Implementation of regional plans is governed by regulatory frameworks, but many of them are still in development, however in s 62 and s 18 *ALSA* directs that regional plans are valid in their absence.

83. Monique Passelac-Ross and Karin Buss, *Water Stewardship in the Lower Athabasca River: Is the Alberta Government Paying Attention to Aboriginal Rights to Water?* (Calgary: Canadian Institute of Resources Law, 2011), 37–40. Online: <http://dspace.ucalgary.ca/bitstream/1880/48638/1/StewardshipOP35w.pdf>; Monique Passelac-Ross, "Public Participation in Alberta's Land-Use Planning Process," *Canadian Institute of Resources Law* 112 (2011); Nigel Bankes, "The Proof of the Pudding: ALSA and the Draft Lower Athabasca Regional Plan" (Ablawg.ca, April 15, 2011); Parastoo Emami, *Evaluating Procedural Justice in Regional Planning Process: Lessons From Alberta's Regional Plans* (MA thesis: University of Lethbridge, 2014), online at <https://www.uleth.ca/dspace/bitstream/handle/10133/3626/Emami-Parastoo-MA-2014.pdf?sequence=1>; and Shaun Fluker, "The Right to Public Participation in Resources and Environmental Decision-Making in Alberta," *Alberta Law Review* 52, no. 3 (2015): 567.

84. South Saskatchewan Regional Plan [SSRP], online at <https://landuse.alberta.ca/LandUse%20 Documents/South%20Saskatchewan%20Regional%20Plan%202014-2024%20-%20February%202017.pdf>.

85. Jaremko *Critical Exploration SSRP*, supra n78. The North Saskatchewan Regional Plan has been started.

86. *Understanding Land Use in Alberta* (2007), 33. Online: <https://landuse.alberta.ca/LandUse%20Documents/ Understanding%20Land%20Use%20in%20Alberta%20-%202007-04.pdf>.

87. SSRP, 27.

88. SSRP, 79, emphasis added.

89. *Paris Accords* (2015), online at <http://unfccc.int/files/essential_background/convention/application/pdf/ english_paris_agreement.pdf>.

90. Article 2 of Paris Accord. For doubters, there is a website called *Skeptical Science* devoted to countering the arguments against global warming: http://www.skepticalscience.com.

91. IPCC, 2014: Climate Change 2014: Synthesis Report. Contribution of Working Groups I, II and III to the Fifth Assessment Report of the Intergovernmental Panel on Climate Change [Core Writing Team, R.K. Pachauri and

L.A. Meyer (eds.)]. IPCC, Geneva, Switzerland, 151 pp. Online: < http://www.ipcc.ch/pdf/assessment-report/ar5/syr/SYR_AR5_FINAL_full_wcover.pdf>, 2–3. Figure SPM.1a. Colours indicate different data sets.

92. Christopher C. Shank and Amy Nixon, *Climate Change Vulnerability of Alberta's Biodiversity: A Preliminary Assessment. Biodiversity Management and Climate Change Adaptation Project* (Edmonton: Alberta Biodiversity Monitoring Institute, Edmonton, 2014) [Shank and Nixon, *Biodiversity and Climate Change*], 1. Online: <http://biodiversityandclimate.abmi.ca/wpcontent/uploads/2015/01/ShankandNixon_2014_ClimateChangeVulnerabilityofAlbertasTerrestrialBiodiversity_ABMI.pdf?>.

93. University of Alberta, Environmental Research and Studies Centre [UA ERC]. Online: <https://sites.ualberta.ca/~ersc/water/climate/impacts4.htm>.

94. Amec 2009, *South Saskatchewan River Basin in Alberta Water Supply Study*, Alberta Agriculture and Rural Development. Lethbridge, Alberta [Amec 2009, SSRB Water Supply Study] online at <http://www1.agric.gov.ab.ca/$Department/deptdocs.nsf/all/irr13053/$FILE/ssrb_main_report.pdf >, 36. It notes that flows were relatively high during the twentieth century and variability is likely to be higher in the future, with climate change studies indicating that "future reductions in flows are more likely than increases in all of the sub-basins," iii.

95. IPCC 2014, supra n91, 8–11.

96. Shank and Nixon, *Biodiversity and Climate Change*, supra n92.

97. Amec 2009, SSRB Water Supply Study, supra n94, 58.

98. Ibid., 57–8.

99. Ibid. That Report predicted that in 2050 Alberta would see future annual precipitation range from a decrease of 3.8 per cent to an increase of 11.5 per cent, with an average increase of 3.6 per cent. Temperature increases ranged from 1.5°C to 2.8°C. Despite the increased precipitation, streamflows were predicted to decrease by 8.4 per cent, averaged across the sub-basins and the various climate models. However, on average, the simulations indicate future reductions in flow in all of the sub-basins of the SSRB.

100. Anita Shepherd, Karen M. Gill, and Stewart B. Rood, "Climate Change and Future Flows of Rocky Mountain Rivers: Converging Forecasts from Empirical Trend Projection and Down-Scaled Global Circulation Modelling," *Hydrological Processes* 24 (2010): 3864–77, 3864.

101. Stefan W. Kienzlea, Michael W. Nemethb, James M. Byrnea, and Ryan J. MacDonalda, "Simulating the Hydrological Impacts of Climate Change in the Upper North Saskatchewan River Basin, Alberta, Canada," *Journal of Hydrology* 412–13 (2012): 76.

102. Mesgana Seyoum Gizaw and Thian Yew Gan, "Possible Impact of Climate Change on Future Extreme Precipitation of the Oldman, Bow and Red Deer River Basins of Alberta," *International Journal of Climatology* 36 (2016): 208.

103. UA ERC, 7. Online at < https://sites.ualberta.ca/~ersc/water/climate/impacts7.htm>.

104. Amec Report, Water Storage Opportunities in The South Saskatchewan River Basin in Alberta (2014) online at <http://www1.agric.gov.ab.ca/$department/deptdocs.nsf/all/irr15015/$file/ssrb-water-storage-main-report.pdf?OpenElement>.

105. Amec 2009, SSRB Water Supply Study, supra n94, Executive Summary, vi.

106. Arlene Kwasniak, *Climate Change and Water: Law and Policy Options for Alberta* (Calgary: Canadian Institute of Resources Law, 2017). Online: < https://www.cirl.ca/files/cirl/water-and-climate-change-occasional-paper-57.pdf>.

Forest Lawn Lift Station, photo: Sans façon.

watershed+
rethinking public art

ciara mckeown

The Bow and Elbow rivers are deeply intertwined with Calgary's identity and sense of place. Calgarians know the rivers and the watershed recreationally, and have an appreciation of and respect for the rivers and the watershed system by proximity. They are drawn to the water regularly to walk alongside it via hundreds of kilometres of river pathways, to swim, wade, canoe, and paddleboard. There is a visceral, tacit knowledge of these waters. Watershed+, a Calgary public art project, reveals how the rivers and the watershed system are more than a geographical site; they are part of how people build individual and collective identity in this city, and forge a sense of place. The rivers hold people's memories, and shape their experiences. Watershed+ aims to unearth these connections and since 2011, countless city staff and artists have worked together to develop a process that inspires civic and citizen stewardship for the watershed's ongoing health while producing challenging public art for the urban realm.

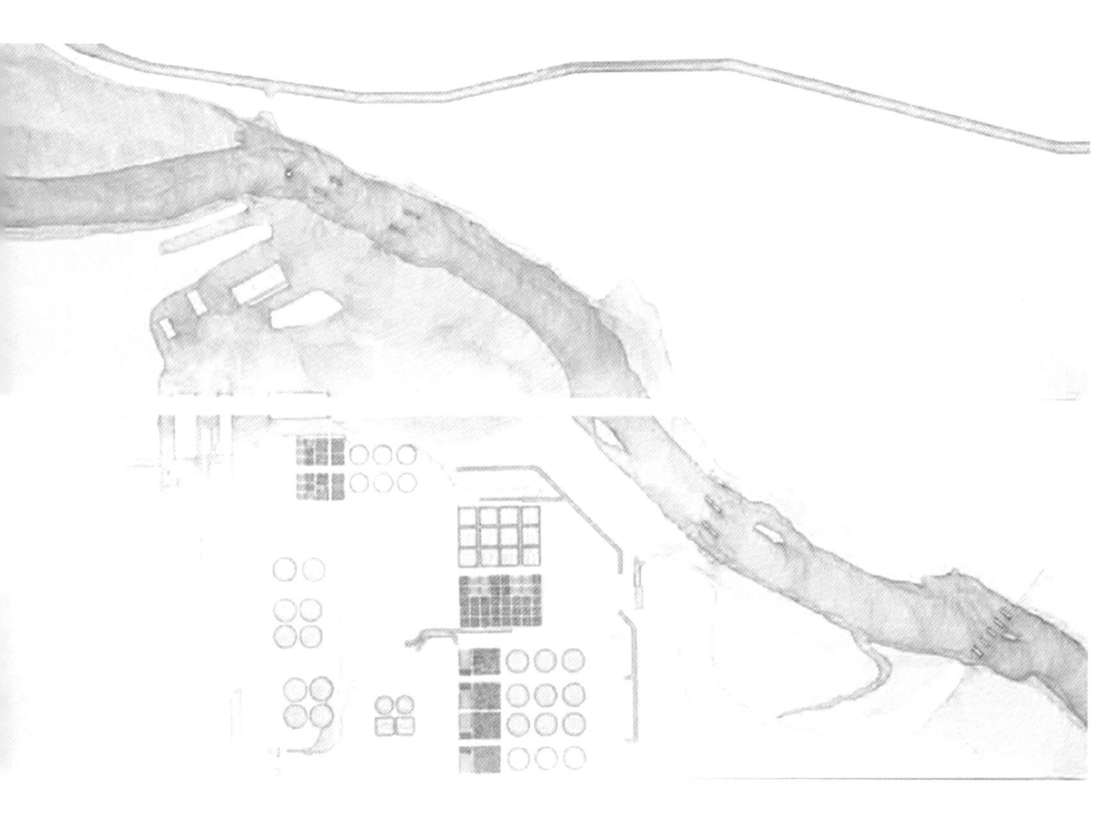

Bow Flow by Rachel Duckhouse, courtesy of The City of Calgary.

A large vinyl Watershed+ sign with bright pink lettering hangs in a studio on the second floor of the City of Calgary Water Centre just southeast of the downtown core. After years of having lead artists embedded within the Utilities and Environmental Protection (UEP) municipal department, they have become such a part of everyday life in the UEP that, save the sign, they otherwise go unnoticed. A rare and visionary initiative in 2009, Watershed+ stems from a philosophical belief that artists, public art, and multidisciplinary collaboration "can create remarkable places that encourage sustainability and stewardship of the environment."[1] The UEP Public Art Program posits that artists are integral to the generation of ideas, with regards to the entire process of evaluating, protecting, and managing the watershed.

After the city of Calgary's Public Art Policy was approved in 2004, UEP and the Public Art Program envisioned the opportunity to establish a new way of working over time, a vision that would be holistic rather than reactive. A ten-year plan was developed to direct public art funds in a strategic and cohesive way to support a diversity of artistic practice and approaches. *A Public Art Plan for the Expressive Potential of Utility Infrastructure* (2007) focused on two of the four business units within the department, Water Services and Water Resources, whose work includes monitoring flow levels and studying flow dynamics; building and upgrading massive water treatment plants that manage drinking, waste, and storm water; and constructing smaller in-situ bank reinforcements to support healthy habitats. This endeavour was rare for its time: few municipalities were developing public art plans specifically for one department, especially one that focused on a subject with both abstract and concrete geographical boundaries, with infrastructure often underground and unseen. In identifying the need for a framework that would support a series of conceptually related artworks over time, rather than just site by site, UEP and Public Art staff helped lay the groundwork for an ongoing collaborative investigation into the watershed system by artists, city staff, and Calgary's communities.

One of the initiatives in the plan was the *Visual Language Project*, identified as the "cornerstone commission" for the UEP public art program that would "create a conceptual framework and visual tone for how UEP wants citizens to recognize and respond to its infrastructure."[2] A collaborative team led by artists Sans façon (Tristan Surtees and Charles Blanc), and made up of an architect, a water engineer, artists, a graphic designer, and a social geographer, responded to this call. Over a twelve-month research period, they began to see that there was much to unearth about how Calgarians understand and relate to the watershed. This would prove to be complex and intriguing material for Sans façon and the interdisciplinary team to delve into. Ultimately, working with city staff across multiple departments, they created Watershed+. At the project's core is Calgary's watershed, which — along with the Water Centre and the spaces, locations, and sites where water staff work, plan, and manage the watershed — forms the whole context. Watershed+, both in its name and in its collaborative process, recognizes that "the art responds to the subject, the art is not the subject."[3] Given the expanse and mysterious force of the watershed it is a vast subject, and such a calling requires a sustained dialogue and collaborative relationship between artists and city staff.

The principles, or "Statements of Belief," laid out in the Watershed+ manual convey the project's philosophical purpose. These beliefs guide, ground, and check all the decisions Watershed+ undertakes. In fact the entire Watershed+ manual, written at the outset to frame a means of moving forward, articulates a purposeful approach that aims to achieve collaboration in new ways. Successful collaborations allow for time — a key factor in making room for disruptions, alterations, and failures. A pilot period protected space for experimentation, allotting time to test projects and initiatives that acknowledged the process. These initial pilot projects were then intended to inform the future development of Watershed+. Time, the desire to be responsive, and a collaborative methodology — an intangible way of working — form the structure from which each project emerges. Thus each project reveals aspects of UEP's complex work, and proposes a reimagining of how citizens might enter into an enriched emotional relationship with the watershed. And because projects stem from this rich process, those who experience it reap a depth of insight, expertise, and the experiences of multiple people, raising the artwork's critical value for citizens, staff, and artists.

One early work, *Fire Hydrant Drinking Fountains* (2012), brings to life how Calgary's invisible water system functions, and reinvigorates the user and casual observer's perception of this by bringing people together socially around water. The work is a series of three differently interconnected shiny bronze pipe fittings and drinking fountains attached to a fire hydrant. *Strangers* brings two people too closely together; *Family* is a set of fountains stepped at incremental heights so everyone can reach; *Group* brings everyone together in varied arrangements; and the overflow runs into dog bowls. Each fountain invokes a curiosity about where our drinking water is coming from, about how it travels and its quality, and each asks us to share in that experience together in different ways.

The project is emblematic of the Watershed+ vision and process. Staff invited the lead artists to be part of thinking through how a water fountain could be a vehicle for conveying information about the city's drinking water system. After researching and conversing with staff, lead artists returned with concepts that "opened up" the inner workings of where water comes from, moving beyond an informational decal. The foun-

Fire Hydrant Drinking Fountain, photo: Sans façon.

tains invite the public at events throughout the city to experience the delivery of our most basic resource. *Fire Hydrant Drinking Fountains* is indicative of how as trust builds over time, and as generous space is given, the ongoing relationship between embedded artists and city staff deepens the meaning of the work, which fosters a shared aspiration for a healthy water ecosystem and an engaged civic dialogue that supports it.

Watershed+ approaches public art by way of process, a long-term durational exercise, where each project and initiative builds upon the last, just as the watershed, with its constant shifts up and down, requires an evolving response. The project did not intentionally set out to challenge existing notions of public art, but did so by being embedded within the bureaucratic system. The artist was there to offer perspective, to add to the thinking and play a complementary role with city staff — not to problem-solve or fill a gap, as is common with traditional public art approaches. As Chris Manderson, the South Region manager for City of Calgary's parks and urban conservation recalls, "the involvement of artists on a project creates the space for a different discussion ...We moved on from asking 'do we want public art on this project' to 'is it a project where we want an artist at the table?'"[4]

sans façon as lead artist

In the formation of Watershed+, it was identified that lead artists embedded within the UEP department over time would bring continuity, criticality, and a unique perspective to the everyday workings of the department. The role of the lead artist, as critic and curator, is more common overseas than in North America, and is important in considering how Watershed+ evolved. As lead artists San façon engendered a critical curatorial process within a municipal department that constantly reconsiders the water system and the work of water management in context.

It takes a certain kind of artist to work within a bureaucratic system, to move from the studio to the cubicle. Sans façon's artistic practice is based on building relationships between people and place. Their approach is to collaborate with others to think through ideas, situations, and definitions of place. They situate themselves in a specific context and take time to meaningfully understand the perspective of this place: what it is and how it is functioning. The intention is not to remake place, but to know the materiality and psychology of it in a fuller way, in order to create a critical, context-specific response. The work is both the process and the end result, depending on what is required. The artist works from within the system, looking at it from a place of knowledge. For Sans façon, the context means more than just site. It is, at any given moment, all of the intricacies, stories, and invisibilities that happen and form the identity of that place. This context-specific methodology has roots in the conceptual art practices of the recent past. These artists become researchers, investigators, facilitators, and magicians. They are "lyhörds," a Swedish term meaning to be an involved listener. Listening, as Magdalena Malm of *Curating Context* notes, "is a key quality in contextual practices, because these artworks are ... in some way porous, reflective — mirroring the shifting images of their surrounding."[5] The nuance lies in how the artists uncover this functioning (people + place), which is constantly changing. Thus, their response shares an experience that opens up the streets and spaces to invite anyone in, an effect that is lasting for those people and that place.

Collaboration is lived at every level of Watershed+, with participants developing on the other's strengths so that every encounter, conversation, and project moves the dialogue forward. Consistent meetings over months and years bear thoughtful, critical results. Sans façon, as lead artists, also became ambassadors for water, for a way of working, and for the mission of collaboration — helping to shape, with city staff, the identity of Watershed+ publicly. Knowing that an accepted methodology and philosophy is the goal, less than a certain end point, the whole Watershed+ team embraces a state of flux, held together by trust in the process. The character of those in the position of lead artist therefore must help facilitate this trust through a demonstration of calm and committed selflessness, which results in a dialogical artistic approach that, as part of its concept, negotiates and intervenes in bureaucratic structures.

Part of that intervention was to have the lead artists experiment with different ways of working, for example a design-team model. *Forest Lawn Lift Station* is one of many stations scattered throughout Calgary, which if marked on a map, together would visualize one of the paths taken by the water in our watershed system. Lift stations are neutral, functional buildings that blend into the landscape. They pump wastewater from low to high areas so it can continue its gravitational flow on to its next destination, the treatment plant. Atop a knoll in the northeast, Forest Lawn's lift station overlooks beautiful city views. It was ready for replacement, and the UEP and lead artists understood that the involvement of artists from the beginning could shape how this building might recast the role of urban infrastructure. Working within the design team, without any preconceived ideas about what the art might be, the intention was to articulate the building's purpose in order to draw people into the story of how water works, to heighten awareness of how we are part of the water system, and to grow an appreciation for this infrastructure. The project is visually dramatic; it evokes curiosity. A map, comprised of LED lights that are connected to sensors monitoring the flow in the pipes, is seen through the building envelope of perforated, dark metal cladding, and represents the exact, scaled representation of the pipes that connect the neighbourhood to the lift station.[6] The changing colours of the lights, in real time, show the happenings inside the lift station at all times. The lead artists and staff on this project, together, demonstrated the creative thinking around how these systems can be understood and appreciated; the project is an example of how Watershed+ is greater than the sum of its parts.

bringing artists in

The intention of the Watershed+ residencies was to introduce national and international artists to the work of UEP, to "support, promote and facilitate innovative and collaborative contemporary art practices responding to issues relevant to Calgary's watershed and water management,"[7] to think through the work of water within a municipal setting, to share their perspectives, and to learn from UEP. This internal access requires delicacy at every level, as well as a thoughtful and careful approach. Artists were supported with time, studio space, and a team of experts with whom they worked in a reciprocal way. Artists had to be willing to come to UEP without a pre-determined approach or desired outcome, willing to have their ideas challenged and deepened by the multiple inputs of others. This often meant that Watershed+ artists had a practice not necessarily grounded in one medium, but rather in process, and wanted to investigate and collaborate to create an informed response.

Rachel Duckhouse, the first artist-in-residence with Watershed+ (2012–13), is a visual artist based in Glasgow. Working in a range of media including drawing and printmaking, her work is formed by exploring complex patterns and systems in nature, human behaviour, and the built environment. During her residency, Rachel became interested in flow dynamics and began a process of visualizing the patterns of flow and the movement of water. Her time in Calgary was particularly poignant as she experienced Calgary's watershed both before and after the 2013 flood. Rachel's research process and large-scale drawings, exhibited publicly in partnership with the artist-run centre TRUCK, articulated the dramatic shift in water flow during this time, and helped Calgarians see the Bow River as two rivers, before and after, to re-imagine it visually in a new way. This residency set the tone for subsequent ones; as a generous, curious artist, Rachel was fascinated by the knowledge and expertise of city staff. She built strong and meaningful relationships with them, and worked in tandem with one engineer to speak about this work and their collaboration at various forums in Calgary. The experiences and approaches of each artist coming in to Watershed+ were shaped and supported by the lead artists and the Watershed+ team. The delicate balance of staff time, the interests of the artist, the critical context, the ongoing logistics of facilitation and project management, are all part of the hidden magic of UEP public art: two lead artists, a public art project manager, coordinator, core group of invested, committed water and public art staff, a communications advisor, and many more, all making this initiative happen, every day.

conclusion

Beyond a series of public art projects, Watershed+ is an exemplar of how the relationships and interactions people have in their daily lives can reveal deep social insights when time is spent examining them. Watershed+'s framework, which might be characterized as "environmental psychology meets conceptual art and relational practice," is a critical investigation that fundamentally shifts definitions and ideas about what public art is, and what the role of the artist can be. There is not enough space left in commissioning public art, in working with artists in a public context; a freedom to move and shift within strict parameters, to find a truer expression of the artist's vision, is lacking. For too long, the field of public art has been weighed down by perpetual sameness; commissioning all the time, in the same way, results in public art that is broadly about objects put into a space. Watershed+ brings us back into the world by asking us to observe and question how we relate to it, and to one another, through art. Just as Watershed+ reveals how integral the often-invisible water infrastructure is to the urban environment, it also reveals how an integrated relationship between artists, city staff, and citizens together in this project quietly furthers emotional connections to and respect for the watershed. Watershed+ demonstrates how trust and reciprocity can reformulate public art beyond itself, how it can truly draw people closer to each other and cultivate a sense of place on an ongoing and constantly evolving basis.

notes

1. A Public Art Plan for the Expressive Potential of Utility Infrastructure, prepared for the Utilities and Environmental Protection Department, City of Calgary, 2007, 2.
2. Ibid., 1.
3. Plus A Succession Plan for Watershed+. Calgary, 2017, 173.
4. Ibid., 174.
5. Magdalena Malm, ed., Curating Context (Stockholm: Public Art Agency Sweden, Art and Theory Publishing, Stockholm, 2017), 12.
6. Plus A Succession Plan for Watershed+. Calgary, 2017, 60.
7. Ibid., 73.

BLACKFOOT CROSSING, BOW RIVI

-From a Sketch by General Strange, R.A.

1882

Blackfoot Crossing, Bow River, N.W.T. by Thomas Bland Strange, 1882, engraving on paper, Glenbow Archives, 61 32.21; Calgary, AB.

swimming in systems

josée méthot & amy spark

alberta ecoTrust

In a commencement address at Kenyon College in Ohio in 2005, author David Foster Wallace began his speech with a joke:

> *There are these two young fish swimming along, and they happen to meet an older fish swimming the other way, who nods at them and says, "Morning, boys, how's the water?" And the two young fish swim on for a bit, and then eventually one of them looks over at the other and goes, "What the hell is water?"*

He goes on to say:

> *If you're worried that I plan to present myself here as the wise, older fish explaining what water is to you young fish, please don't be. I am not the wise old fish. The point of the fish story is merely that the most obvious, important realities are often the ones that are hardest to see and talk about.*

This brings us to the important (and changing) realities of water in Alberta. Just what are we swimming in?

We've seen the headlines: declining snowpacks; the specter of drought and the untameable power of floods; floodplains being mined for gravel; longstanding drinking water advisories on First Nations reserves. The 2017 CIH Community Forum helped highlight some of these issues and the need for action.

Many groups in Alberta are working tirelessly to address complex water issues. They include Alberta's Watershed Planning and Advisory Councils (WPACs) located in eleven watersheds across the province. These place-based groups bring together people from across sectors to research, educate, and find collaborative solutions to regional water issues. They are basically Swiss Army knives for watershed protection, and act as forums to bring different ideas and voices together.

Josée works for the Red Deer River Watershed Alliance, a WPAC working to protect the Red Deer River that flows from Banff National Park eastward through Sundre, Red Deer, and Drumheller. The Alliance actually works to protect all the 50,000 km² area of land that contributes water to the river — the "watershed." This area is home to 300,000 people and spans the ancestral and traditional territories of Treaty 6 and Treaty 7 First Nations.

It can be overwhelming to work on water issues. The urgency and complexity of the issues we face is daunting. Water is an issue connected to many others — the type that forces us to confront our land use, energy, food, and lifestyle choices. While the twentieth century saw major breakthroughs in municipal water management and ushered in an era of safe, cheap, and abundant water (for many, not all), we are now faced with a next generation of challenges. A recent assessment of the health of Canada's freshwaters by WWF-Canada found that several of Canada's most "at-risk" watersheds are in Alberta, facing risks including pollution, habitat loss, and the overuse of water (WWF-Canada, 2017).

Human ingenuity in the twentieth century may have sent man to the moon, but our challenges now are more complex than building a rocket ship. The changing reality of water in the twenty-first century is poised to reshape how we live, how we work, and how we value water. How can we prepare? Lessons from the fields of ecology and resilience thinking suggest we need to think systemically, work adaptively, and broaden participation in order to deal with changing social and ecological systems (Stockholm Resilience Centre, 2015).

This is exactly the type of approach that the Red Deer River Watershed Alliance and Alberta Ecotrust have been exploring in partnership since 2015, through *Project Blue Thumb: Action on Water Quality Issues*. Project Blue Thumb is Alberta's first "social innovation lab" dedicated to addressing water issues, and brings together a diverse team of people to discuss watershed issues and take creative action. The lab team has met numerous times since 2015, and continues to engage individuals from government, industry, the non-profit sector, academia, and the public.

The key water events that make newspaper headlines, such as floods, algae blooms, and invasive species, do not tell the whole story. In order to understand Alberta's water system and the broader systems we are swimming in, a deeper understanding is required. Underlying these events are key trends (e.g., loss of wetland cover), structures (e.g., policies and institutions), and mental models (e.g., the values and attitudes that shape our complex behaviours). Project Blue Thumb came about as a way to look at these deeper drivers of watershed change — they may lurk below the surface, but addressing the structures and mental models that drive system behaviour is key. Our goal has been to find real levers of change in the complex space of watershed management — i.e., how can we make a bigger difference? Our hunch is that we need to work systemically, creatively, and with a broader set of players.

Project Blue Thumb's approach is rooted in the field of social innovation. We look for ways to build knowledge, networks, and skills with key community members, while maintaining a focus on action. Admittedly, the idea of "innovation" can be polarizing — naysayers decry it as a buzzword while promoters have bought into the hype. For us, social innovation is simply about asking: *is this enough? How can we do better?* It is not about inventing some flashy gadget or app that will magically make things better. Instead, it is about reorienting shared efforts around a more systemic perspective, such that we can work together to address deeper leverage points for change. From there, we need to employ a mix of approaches — from the tried and true to the experimental and iterative.

Moving forward, we are reassured in knowing that there are many people working to improve watershed management across the province — including players from First Nations, the non-profit sector, government, industry, academia, the arts, and more. We are, however, compelled to do more. We ask this growing community: do we have the courage to imagine vastly different water futures; to question and dismantle outdated policies and processes; to break free of the idea that somehow our best laid strategic plans will save us? Can we move beyond the idea that mere infrastructural fixes will be sufficient?

In the past decade, urban and local agriculture have boomed in Western Canada. There is a local food transformation afoot. Can we now imagine a future with a parallel local water transformation — one rooted in a people-powered water movement? A future where it is cool to love a wetland, to fish healthy trout, and where everyone knows where their water comes from?

Returning to David Foster Wallace's story, if "*the most obvious, important realities are often the ones that are hardest to see and talk about*," then we have to face Alberta's changing water realities head on.

Morning, Lake Louise, by Frederic Marlett Bell-Smith, 1889, watercolour cn paper, Glenbow Archives, 59.34.7, Calgary, AB.

Shelley Ouellet, **Wish You Were Here . . . (Lake Louise)**, 2001, photograph by M. N. Hutchinson.

Collection of Nickle Galleries, purchased with the support of the Canada Council for the Arts Acquisition Assistance program.

Parce qu'elle est l'eau
soeur de la sainte lumièr
Sur qui flotta l'Esprit
de l'aurore première

— Louis Mercier

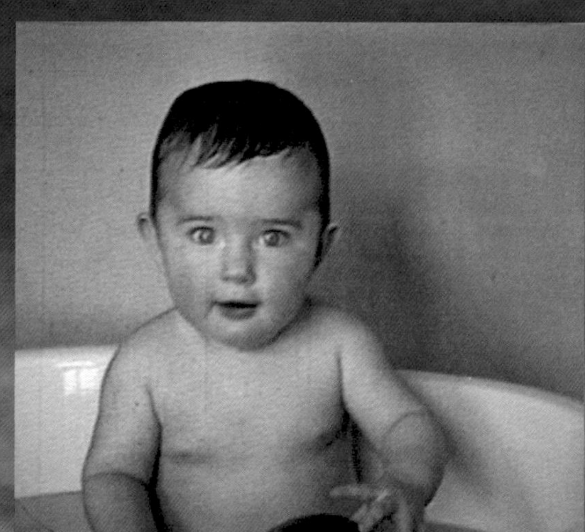

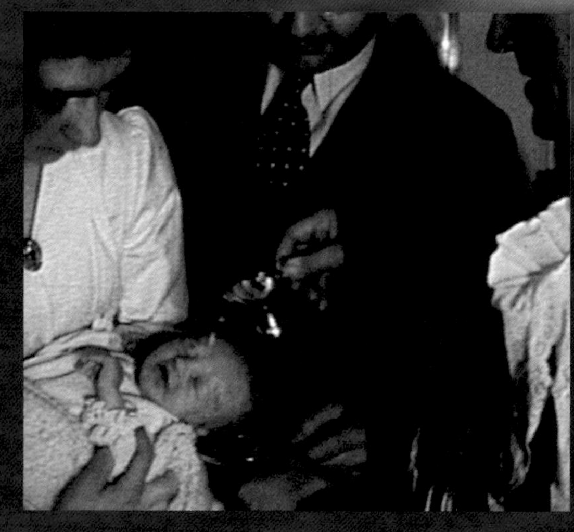

Frame captures from **Gloire à l'eau** (Albert Tessier, ca. 1935/1950) courtesy Bibliothèque et Archives nationales du Québec.

glory to water
gloire à l'eau
by *albert tessier*

(ca. 1935/1950) 16mm, B&W and colour film, silent, 10 minutes

charles tepperman

university of calgary

When the 16mm film format was introduced by Kodak in 1923 it opened up the domain of filmmaking to non-professionals. While home movies made up the largest portion of this activity, many amateurs became proficient filmmakers who used the medium to chronicle their travels, support their professional work, or simply as means of creative expression. Albert Tessier was a priest and amateur filmmaker who used film for all three of these purposes. Based in the St. Maurice River area north of Trois-Rivières, Quebec, Tessier made about seventy short films between 1927 and the 1940s.[1] He used his films to record his canoe trips and also as part of the work he did for the St. Maurice Forest Protective Association and later for the Catholic church's program in family education. His films captured the region's nature, as well the rural life and customs of its inhabitants, often through a poetic and sacred lens. One of Tessier's most acclaimed films was *Gloire à l'eau*, a tribute to the many practical and spiritual uses of water.

Tessier filmed *Gloire à l'eau* in the mid-1930s, supplementing and re-editing it often over the next decade and a half. The print that survives is shorter than some versions of the film and we can see traces of his revisions in the different qualities of film stock (some black and white, some colour) and varying exposure levels.[2] Tessier described the film as "abandoned," indicating some dissatisfaction with its final state. Amateur moviemakers during the 1930s often made poetic and lyrical polished films about nature, so Tessier was part of a broader community of filmmakers in this respect. Tessier described *Gloire a l'eau* as his "favorite movie ... a subject a cineaste should try."[3] Different versions of the film were exhibited widely, and it was included in the "International Amateur Movie Show" presented by Columbia University in New York in April 1938.[4] Though typically Tessier presented the film with music or spoken commentary, the version that survives is silent.

Gloire à l'eau presents an inventory of water's many uses, both spiritual and banal. The first shot of the film shows a silent, colour image of a man (perhaps Tessier) speaking and drinking water from a cup before an intertitle introduces the metaphysical dimensions of film's subject: "Because it is water/ Sister of holy light/ above which floated the spirit/ of the first dawn." We see shots of waves from a high angle and surf crashing against rocks, and the visually dramatic quality of water is juxtaposed against the religious themes of the next intertitles: "Praise be, my Lord/ for sister water/ which is plentiful/ useful and humble/ and precious and pure." The camera pans across water in a landscape setting of a lake or river with hills rising above it.

Following these introductory verses and images, the film presents a series of water types. According to surviving documentation, the film's thesis was to organize water into two broad categories, the first of which was "L'eau, mère de la vie" [Water, Mother of Life]. The first instance of this is "L'eau-sanctification" [Water-Sanctification], explained by Tessier as "Mother of Supernatural Life."[5] Here we see an interior scene with faucets and ceremonial accoutrements, and then a priest performing a baptism, pouring holy water on a baby's head. Second in the film's typology is "L'eau-beauté" [Water-Beauty], which for Tessier is the "Mother of Intellectual Life." Here we see images of water running past plantlife, a swirling pool near some rocks, ripples, and rapids, and then cutting larger rivers and their rapid and gentle flows. The next category is "L'eau-nourriture" [Water-Nourishment], "Mother of Animal Life (Men and Beasts)." We see a moose standing in shallow water drinking, and in the next shot a man drinks with a cup at the water's edge and then we see him casting a fishing-pole, before cutting to fishing boats on larger bodies of water. By editing in this way Tessier employs a categorical form of filmmaking, juxtaposing different users and uses of water. "L'eau-engrais" is next, focusing briefly on instances of "Water-Fertilization" ("Mother of Vegetable Life") and showing large trees along the edge of a river.

The second broad category of water is its practical uses as "Collaboratrice de l'homme" [Man's Collaborator]. This includes "L'eau-force" [Water-Power], and we see a water-driven mill turning, a steam locomotive charging past the camera, and tumultuous rapids churning in the river and passing an electric generation plant, leading from water flow to electrical wires. "L'eau-route" is next, showing the means of transportation provided by water, including canoes passing on a large river, larger boats diverting lumber, and then a mass of logs passing down the river, through a chute, and into rapids. Finally, a sailboat is shown, pointing to more peaceful methods of traveling, and large passenger ships are seen travelling down the river, belching smoke into the air. "L'eau-hygiene" concludes Tessier's typology, as a large pitcher of water is poured into a clean white basin and a nurse bathes a baby with soap and water. Further images compare different kinds of bathing, from babies in the tub, to a man washing his face in the river, and then a large group of children splashing, swimming, and playing in the water. The film's conclusion returns to verses that connect water with prayer and religion: "Praise be, my Lord/ For sister water/ which is plentiful/ useful and humble/ and precious and pure." Graceful ripples of blue water are shown, and are almost abstract in quality; the film's last images are of light reflections on rippling water and the moon.

Gloire à l'eau is an excellent example of thoughtful amateur filmmaking from the 1930s. It presents a compilation of a varied informational and poetic footage of water that is then organized into categories of use and spiritual significance. In this way, Tessier could employ his film as both an expression of individual creativity and as a demonstration of religious ideas, showing how the sacred and beautiful qualities of water permeated all aspects of life. Over the course of the 1930s, amateurs developed networks of movie clubs to support their activities, and national and international organizations emerged to help coordinate efforts and circulate their films. After 1941, some of Tessier's films (including *Gloire à l'eau*) were distributed by Quebec's government cinema service.[6] Today, Tessier is best remembered as the namesake for the Prix Albert-Tessier, awarded each year to an individual for their outstanding career in Quebec cinema.

notes

1. René Bouchard, *Filmographie d'Albert Tessier* (Montréal: Les Éditions du Boréal Express, 1973).
2. The film is preserved by Bibliothèque et Archives nationales du Québec.
3. Bouchard, 55. All translations by author.
4. "Three Cinematographer Awards On International Show Program," *American Cinematographer*, April 1938, 170.
5. *Filmographie*, 55.
6. Louis Pelletier, "Un cinéma officiel amateur : les racines artisanales du cinéma gouvernemental québécois," in *L'amateur en cinéma – Un autre paradigme*, Valérie Vignaux and Benoît Turquety (eds.) (Paris: AFRHC, 2016).

water rights/ water justice

adrian parr

Adopted in November 2002, Article I.1 of *General Comment 15* that the United Nations Committee on Economic, Social, and Cultural Rights states: "The human right to water is indispensable for leading a life in human dignity." Similarly, water and sanitation are the focus of UN Sustainable Development Goal 6. With it the organization aspires to "ensure availability and sustainable management of water and sanitation for all."[1] Then we have *Resolution 64/292*, which was implemented by the United Nations General Assembly on July 28, 2010. It recognizes "the right to safe and clean drinking water and sanitation as a human right that is essential for the full enjoyment of life and all human rights."[2]

For the United Nations General Assembly, realizing the human right to clean and safe water and sanitation requires that "states and international organizations ... provide financial resources, capacity-building and technology transfer, through international assistance and cooperation, in particular to developing countries, in order to scale up efforts to provide safe, clean, accessible and affordable drinking water and sanitation for all."[3] There are a few basic water and sanitation statistics that are often cited by international or non-profit organizations to support and advance Sustainable Development Goal 6. As of 2015 there were approximately 664 million people, over half of whom were in Sub-Saharan Africa, who used water from sources that were not protected from outside contamination, otherwise referred to as an "unimproved water source."[4] In 2015, despite 4.9 billion people benefiting from the improved sanitation facility, 2.4 billion people were not.[5] Unmanaged fecal waste continues to pose a serious risk to the health and well-being of many people, the majority of whom live in the growing slum settlements of the developing world. If currently 827 million people are living in slums and each month 5 million people are added to the slum populations of the developing world, the challenge of accessing clean water and sanitation is only mounting as each day passes.

I am not going to take issue with whether or not water is indispensable for a life of dignity or if it is the basis for realizing other human rights such as health, gender equity, and so on. I am also not going to take issue with the UN's request that states and international organizations act to institute access to clean water and sufficient sanitation services. These points are all axiomatic: without water there is no life on earth. On average, a person can survive approximately three weeks without food and only three days without water. Rather, I am interested in exploring how the realization of a right to water is a political action, one that has the potential to turn into political power. The guiding question here is: what kind of political powers does realizing a right to water in the US produce?

More specifically I will unpack how the demand for environmental justice contains within it an appeal to human rights. In this instance, that would be the right to clean water. More precisely: how does the appeal create the conditions through which a political subject emerges? Remember, the United Nations Resolution 64/292 explicitly connects clean drinking water and sanitation to the "realization of all human rights."[6] Given this, is it politically effective to inscribe water politics with a rights-based approach? Can such an inscription be the basis for an inclusive and emancipatory political project?

water wealth and the US

What the commonly cited statistics of the developing world conceal from view are the water and sanitation trials and tribulations plaguing communities of the "developed world," such as the United States. The water crisis at Flint, Michigan is just one example that broke through the pervasive silence surrounding water injustice for one of the wealthiest countries in the world. In 2016 the United States was ranked the thirteenth richest country in the world (Qatar was first). The ranking was calculated by using Gross Domestic Product (GDP) based upon per capita purchasing power parity (PPP).[7] Like most statistics, this one needs some context if we want to comprehensively understand how the wealth-inequity nexus amplifies the increasing urgency of clean water practices and conditions in the US.

In addition to being one of the wealthiest countries, the USA is also one of the most inequitable. Wealth is concentrated in the hands of a few and is not evenly distributed — or if we want to follow neoliberal parlance, it does not trickle down.[8] In 2011, the Center on Poverty and Inequity at Stanford University sounded an alarm bell, reporting that the overall level of wage inequality in the United States was fast "approaching the extreme level that prevailed prior to the Great Depression."[9] The average annual income of the top 1 per cent is $1,153,293 as compared to the average income of $45,567 for the remaining 99 per cent.[10] Put simply, in the US the top 1 per cent earns 25.3 times more than the remaining 99 per cent.[11]

However, it is not only income disparities that flesh out the picture of inequity in the US. In 2016, 549,928 people on any given night were homeless; 22 per cent of these were children, and 69 per cent were under the age of 24 years.[12] In 2009 US students living in families from the bottom 20 per cent of family incomes were five times more likely to drop out of school than students in high-income families (top 20 per cent of family incomes).[13] Employment success and income earning capacity are directly connected to the number of educational qualifications a person completes. Yet, it is not only the poverty individual families experience that impacts high school dropout rates. Some communities struggle with multiple forms of disadvantage — under-resourced schools, insufficient number of after-school programs, few parks and children's playgrounds, a lack of healthy food options, poor health, reliance on welfare, substance abuse, unemployment, poor quality housing, residential instability, overcrowded housing, and polluted environments, all of which materialize into numerous forms of violence (negative stereotyping, racial profiling by law enforcement, segregation, crime, and discrimination).[14]

Inequity is experienced differently according to where a person lives, their gender, age, sexuality, ability, race, and ethnicity. There are stark gaps between who has access to quality healthcare, education, and housing and who does not. These gaps deepen and complicate how inequity works across the US and it is typically poor minorities who suffer the most. As William Julius Wilson has pointed out, any comprehensive understanding of how poverty works must combine structural explanations with cultural ones.[15] Nevertheless, as the recent water crisis in Flint highlighted, not only is inequity shaped by the intersection of structural and cultural forces, it is also influenced by a third hardship variable: environmental adversity. A study by Mona Hanna-Attisha et al. analyzed elevated lead levels in children from Flint who were associated with the water crisis. The authors highlight not only policy failure but also a correlation between "socioeconomically disadvantaged neighborhoods" and the "greatest elevated blood lead levels."[16]

One might argue that the case of Flint is less an instance of environmental hardship than it is a structural issue. Namely, it is the result of governance failure, or insufficient water infrastructure. Carla Campbell et al. correctly identified unsuccessful water governance as a key factor in the Flint water crisis. The authors point a clear finger of blame in the direction of the city of Flint, which failed to "properly treat its municipal water system."[17] Pushing back against Hanna-Attisha et al. and the view that Flint's water crisis was an "aberration — a single policy failure," Michael Greenberg has justifiably insisted on the importance of developing critical infrastructures that reliably deliver clean potable water and advance public health.[18] Critical infrastructure, he explains, "is a term used to identify public and private assets that are required for society and the economy to function."[19] He notes that many poor US neighbourhoods, like the one in Flint, suffer from "relatively high burdens of environmental deterioration that includes water and other infrastructure systems."[20] The Environmental Protection Agency (EPA) estimates that the cost to repair and replace aging US water infrastructure over a twenty-five year period will be in the vicinity of $1 trillion.[21]

Greenberg argues that in order for governments to ensure the protection of human health and safety, more comprehensive risk assessments need to be conducted on all US water sources and infrastructures. Water infrastructure is without doubt capital intensive. That is a hard call in the current climate, where funding to the Environmental Protection Agency is under attack by a political administration hostile to government regulation. For this reason Greenberg suggests that public health officials are well positioned to pressure "elected officials and their administrative staff" to provide "safe water distribution" to communities. He defends his position, stating that such distribution is not only "essential" to ensuring public health, it is also "environmentally just."[22] This is a politically strategic move on his behalf, aimed at moving the dial on an urgent situation. In effect, though, his argument flattens the political terrain.

In assigning public health officials the role of representing the water needs of underprivileged groups within the field of institutionalized politics, the work of the public health official moves beyond communicating the public health concerns of a specific community to holding political power. In this situation a public health official holds a representative power intended to be responsive to ensuring that public health needs are met. Operating within the system of a liberal democracy, the public health official speaks out against an environmental injustice that it is manifesting itself in poor public health outcomes, such as high lead levels in African American children living in underprivileged neighbourhoods. The public health official is tasked with working together with elected officials to realize health reform, using the mechanisms of policy and legislation. Greenberg's position relies upon a liberal assumption that in a US democracy people effectively participate in government through a peaceful system of political representation that distributes political power amongst various representatives elected to office by US citizens of voting age. The system of representative democracy relies on achieving general consensus, and political change occurs in increments. Political action takes place along a horizontal plane, which is translated into political change through mechanisms of reform that take place along the vertical axis of institutionalized politics. The assumptions of reform-based political change, however, leave intact the undemocratic systems that led to the water crisis in the first place.

In 2011 the city of Flint was declared to be in a financial emergency, prompting officials to seek out cheaper water supply alternatives. It eventually decided in 2013 to switch the city's water supply from Detroit Water and Sewerage Department (DWSD) to the Karegnondi Water Authority (KWA). The projected cost savings to the city were estimated in the vicinity of $200 million over a twenty-five-year period.[23] To do this the city had to build a pipeline to connect to the DWSD. After the announcement was made, Detroit Water put the city on notice stating that it would stop water supplies within a year. This meant the city needed an interim source of water while it built the water pipeline. On April 25, 2014, Flint began drawing its water source from the Flint River — ignoring advice to use corrosion inhibitors to ensure the water didn't decay the water pipes. In a press release, city officials ignored the warnings, dispelling them as a "myth"; then-Mayor Dayne Walling reassured residents that the Flint River water source was safe, and described it as "regular, good, pure drinking water."[24]

When residents started complaining a month later about the odour and colour of the water, Michigan Department of Environmental Quality engineer Michael Prysby stated although there were differences in the new water's hardness, smell, and taste, it was nonetheless safe.[25] By August 2014 Flint officials tested the water in the "boil zone." Their tests brought up traces of fecal coliform bacteria and E. coli. The city blamed the results on an "irregular E. coli test result,"[26] and responded by issuing a boil water notice to the west side of Flint and increasing the amount of chlorine in the water.[27] By January 2015 the amount of total trihalomethanes (TTHM) reached dangerous levels and the city was found to be in violation of the Safe Drinking Water Act. On January 12, 2015 Detroit Water and Sewerage Department offered to reconnect water to Flint and waiver the $4 million reconnection fee; however, concerned about higher water costs ($12 million a year), Flint officials rejected the offer.[28]

Late in 2015 a Virginia Tech research team tested Flint's water supply and found a lead content of 13,200 parts per billion.[29] According to the EPA a safe maximum lead content is fifteen parts per billion. Water is considered a hazardous waste when its lead content reaches 5,000 parts per billion. With pressure mounting, the city issued a public statement in September 2014 clarifying and reassuring the public that it was in full compliance with the Federal Safe Drinking Water Act. Governor Snyder's chief of staff, Dennis Muchmore, went so far as to question the integrity of those who raised the alarm bell on Flint's water supply; downplaying the seriousness of the situation he maintained that instances of child lead poisoning were being used as a "political football."[30]

By January 8, 2016, Flint residents protested on the lawn of Flint's City Hall, demanding that state officials be held accountable for water poisoning. In response to the Flint water crisis President Obama signed a Michigan Emergency Declaration on January 16, 2016, releasing $5 million in federal aid to alleviate the public health crisis. On January 21, 2016, the EPA delivered an emergency order to the city and water filters were issued to residents.[31] A year later the people of Flint still couldn't drink their water without using a filter, and a great number of families still depended upon bottled water.[32]

There were many moments along the way in the Flint water crisis where different forms of civic participation were overlooked and overridden altogether. Numerous different warnings about Flint's water quality came from concerned residents, public officials, and water researchers. Flint is a majority African American city where 40 per cent of the citizens live in poverty. The Michigan Civil Rights Commission determined that "structural and systemic racism combined with implicit bias led to decisions, actions, and consequences in Flint" that would not have occurred in white communities such as Ann Arbor or Birmingham.[33]

What Flint exposed was the role that bias and structural racism played in providing water services to Flint residents. In addition to this, putting a price tag on a public good such as water in a climate of financial emergency compounded the problem, as risky cost-cutting measures further distorted the institutional forms of racism that the Michigan Civil Rights Commission identified. In this context, the intersection of poverty, racism, low public financial resources, environmental injustices, and health intersected to form a punitive and dangerous combination that rendered reform-based political change dysfunctional. Even one year later residents in Flint were still using filters and bottled water, and they were instructed to run their water faucets longer in order to flush out bacteria.

What really turned the political dial on the Flint water crisis were protests by residents, and an active culture built around "citizen science," whereby everyday people began to participate in collecting important water samples that would be used to develop a case. The Virginia Tech Research Team of the Flint Water Study built that case. Turning public health officials into the representatives of a political struggle that is largely occurring at the intersection of social, economic, cultural, and environmental disadvantages shifts the unit of politics away from the unpredictable horizontal plane of collective action and protest toward predictable vertical activities such as policy reform and the development of critical infrastructures. Put differently, revolutionary rupture was liberalized as incremental, reform-based change. Admittedly, not all revolutionary struggles are equal, not all advance the same agenda, and in order to be politically effective such struggles must at some point be translated into institutional form and organized political action. For these reasons, I understand the politics of water to rest upon a common right — not an individual human right to water but rather a right that recognizes multiple sites of struggle within water politics.

rights-based theory

While it may seem obvious to state all people have a basic right to clean water and sanitation the deeper problem of how to enforce that right persists. To delink rights-based discourse from the social, economic, cultural, and political struggles in which any right is realized depoliticizes how water resources work in contemporary life. In the current neoliberal climate, the privatization of water resources is the new norm dominating the realization of water rights all over the world. A rights-based framework assumes the connection between water resources and human well-being is a property relation. As the UNHCR states: "Human rights are rights that are inherent to all human beings."[34] Furthermore, a human right is inalienable and universal that is "guaranteed by law."[35] The right to water appeals to a distinct and bounded entity demarcated and protected under the law. This reinforces the very nature of the problem we face vis-à-vis water. Namely, by understanding water as a system of flows that exceed the artificial boundaries human property relations impose on it.

All in all, identifying water as a human right re-represents the sociality of water relations using a neoliberal worldview. Water is a property that one "owns." This ultimately justifies leaving inequity up to the free market to solve. Unsurprisingly, from 1990 to 2002 the number of people across the world who were served by private water companies increased from 51 million to 300 million.[36] During the same time period, as Gary Wolff and Meena Palaniappan report, six water companies increased their operations in twelve countries to more than fifty-six countries.[37] History teaches us, however, that using free-market mechanisms to address social and environmental burdens leaves intact structures of exploitation and oppression that intensify disadvantage, dispossession, and hardship. As Food and Water Watch have calculated, a privately owned water utility service for the average US household costs 59 per cent more than a public water service (approximately $158 more a year), and when water is privatized rates increase approximately three times the rate of inflation.[38]

Furthermore, the assumption that water relations exist within the legalistic frame of property relations also exposes a disturbingly anthropocentric bias that ignores the central importance of water to the flourishing of ecosystems and other-than-human species. A quick example suffices to describe the interspecies conflict of water struggles.

In response to the ongoing drought in California and the Flint Michigan water crisis, in December 2016, then-US President Obama signed into law the $11 billion *Water Resources Development Act*. In addition to addressing Flint's water crisis, California was assigned $558 million for water projects that ranged from water recycling to desalination plants and flood control initiatives. Before the Senate voted on the new legislation, a "midnight rider," allowing for maximum pumping of the San Francisco Bay Delta, was attached to it. The delta is the largest estuary on the Pacific Coast of America. It is where the south-flowing Sacramento and north-flowing San Joaquin Rivers meet. Pumping of water from the delta is already incurring a heavy toll for the delta's ecosystems, leading to declining fish populations.[39] In this case, realizing the right to water blocks from view the rights of other-than-human species to survive and flourish.

Water infrastructure, as Greenberg understands it, are built structures such as water purification plants, storm water systems, levees, dams, piping systems, and aqueducts. Greenberg speaks of developing crucially important water infrastructures that provide up-to-date and well-managed physical structures that efficiently pump, treat, store, and deliver water to all communities across the United States. I contest his theory that a right to water has been realized if physical water infrastructures are developed. Instead I would suggest that we create infrastructures for new, inclusive social relations through which a common right to water can be actualized.

As a social relation, the struggle for clean water allows us to recognize the importance of working and creating solidarities across social movements. Clean water, as already demonstrated, is not an isolated phenomenon; it is imbricated in issues such as clean energy, minimum wage, access to affordable heath care and quality education, quality housing, healthy food, tolerance of diversity, better resources for public institutions, all of which are central to a variety of other social movements — women's, LGBTQ, migrant, Black Lives Matter, Indigenous, peace, animal rights, citizen science, and climate justice. Although these struggles operate at different scales from the local to the national or even global, from different points of identification (identity politics), or as instances of class warfare, it may be near impossible to unite them into one political class. Nonetheless, their activist energies can be politically powerful if solidarities form across these fronts.

These criticisms aside, to completely toss out rights-based theory when trying to respond to the collective problem of access to clean water would be to throw the baby out with the bathwater. The notion of rights must be collectivized to reflect a right that is meaningless if isolated into an anthropocentric view (i.e., restricted to a human right), and/or a right premised upon individual ownership. All in all, the right to water is not a self-identical right. It is filled with contradictions because as a political project it is never complete. For instance, the water struggle of Flint residents is ongoing.

I understand the right to water to be a common right that is articulated in relationship to history, local conditions, material resources, political failures and successes. A common right, as I understand it, is a transdisciplinary, transspatial, transtemporal, transpecies right that all fairly occupy, including future generations and other-than-human species. As a common right, water coheres with political antagonism and the struggles of the oppressed as they challenge and attempt to transform the violence inherent to a capitalist mode of production and reproduction. The important collaborative role citizen science played in mobilizing political change in Flint shows that realizing a common right to water can institute new forms of collective power, by developing social infrastructures needed for inclusive and emancipatory political projects.

notes

1. United Nations (2017), Progress of Goal 6, accessed December 21, 2017, https://sustainabledevelopment.un.org/sdg6.

2. United Nations General Assembly (2010), Resolution 64/292 (3 August), accessed May 11, 2017, http://www.un.org/es/comun/docs/?symbol=A/RES/64/292&lang=E.

3. Ibid.

4. UNICEF and World Health Organization (2015), "25 Years: Progress on Sanitation and Drinking Water: 2015 Update and MDG Assessment," 11, accessed December 19, 2017, http://files.unicef.org/publications/files/Progress_on_Sanitation_and_Drinking_Water_2015_Update_.pdf.

5. United Nations (2017).

6. United Nations General Assembly (2010).

7. Jonathon Gregson (2018), "The Richest Countries in the World," *Global Finance*, March 1, accessed May 12, 2017, https://www.gfmag.com/global-data/economic-data/richest-countries-in-the-world?page=12.

8. Estelle Sommeiller and Mark Price (2015), "The Unequal States of America: Income Inequality by State 1917–2012," *Economic Policy Institute*, January 26, accessed May 11, 2017, http://www.epi.org/publication/income-inequality-by-state-1917-to-2012/.

9. Center on Poverty and Inequality (2011), "20 Facts About US Inequality Everyone Should Know," Stanford University, January 26, http://inequality.stanford.edu/publications/20-facts-about-us-inequality-everyone-should-know.

10. Sommeiller, Price, and Wazeter (2016).

11. Ibid.

12. Megan Henry, Rian Watt, Lily Rosenthal, and Azim Shivji, Abt Associates (2016), *The 2016 Annual Homeless Assessment Report to Congress*, The US Department of Housing and Urban Development (November), 1, accessed May 11, 2017, https://www.hudexchange.info/resources/documents/2016-AHAR-Part-1.pdf.

13. C. Chapman, J. Laird, N. Ifill, and A. KewalRamani (2011), *Trends in High School Dropout and Completion Rates in the United States: 1972–2009* (NCES 2012-06) (Washington, D.C.: National Center for Education Statistics, Institute of Education Sciences, U.S. Department of Education), accessed May 9, 2017, http://nces.ed.gov/pubsearch/pubsinfo.asp?pubid=2012006.

14. Robert Bullard (2000), *Dumping in Dixie: Race, Class and Environmental Quality*, 3rd ed. (Boulder: Westview Press); Russell Rumberger (2011), *Dropping Out: Why Students Drop Out of High School and What Can Be Done About It* (Cambridge, MA: Harvard University Press); Robert J. Sampson, Jefffrey D. Morenoff, and Thomas Gannon-Rowley (2002), "Assessing 'Neighborhood Effects': Social Processes and New Directions in Research," *Annual Review of Sociology* 28 (August): 443–78; Kristin Turney (2010), "Neighborhood Disadvantage, Residential Stability, and Perceptions of Instrumental Support Among New Mothers," *Journal of Family Issues* 31, no. 4: 499–524.

15. William Julius Wilson (2010), "Why Both Social Structure and Culture Matter in a Holistic Analysis of Inner-City Poverty," *Annals of the American Academy of Political and Social Science* 629, no. 1 (May): 200–19.

16. Mona Hanna-Attisha, Jenny LaChance, Richard Casey-Sadler, and Allison Champney Schnepp (2016), "Elevated Blood Lead Levels in Children Associated with the Flint Drinking Water Crisis: A Spatial Analysis of Risk and Public Health Response," *American Journal of Public Health* 106, no. 2 (February): 283–90.

17. Carla Campbell, Rachael Greenberg, Deepa Mankikar, and Ronald Ross (2016), "A Case Study of Environmental Injustice: The Failure in Flint," *International Journal of Environmental Research and Public Health* 13, no. 10 (October): 1.

18. Michael Greenberg (2016), "Delivering Fresh Water: Critical Infrastructure, Environmental Justice, and Flint, Michigan," *American Journal of Public Health* 106, no. 8 (August): 138–1360.

19. Ibid., 1358.

20. Ibid.

21. United States Environmental Protection Agency (2013), *Drinking Water Infrastructure Needs Survey and Assessment: Fifth Report to Congress* (April): 4, accessed May 10, 2017, https://www.epa.gov/sites/production/files/2015-07/documents/epa816r13006.pdf.

22. Greenberg, 1359.

23. Merrit Kennedy (2016), "Lead-Laced Water in Flint: A Step by Step Look at the Makings of a Crisis," *National Public Radio*, April 20, accessed December 21, 2017, https://www.npr.org/sections/thetwo-way/2016/04/20/465545378/lead-laced-water-in-flint-a-step-by-step-look-at-the-makings-of-a-crisis.

24. Rick Snyder (2016), "State of Michigan Executive Office: Lansing," 16, accessed December 10, 2017, https://www.documentcloud.org/documents/2696071-Snyder-Emails.html#document/p16/a272881.

25. Ron Fonger (2014), "State Says Water Meets All Standards But More Than Twice the Hardness of Lake Water," *Michigan Live*, May 23, accessed December 11, 2017, http://www.mlive.com/news/flint/index.ssf/2014/05/state_says_flint_river_water_m.html.

26. The city stated: "Testing over the last 48 hours has shown that the water tests negative for the presence of fecal coliform bacteria and E. coli, but an abnormal test result triggered the advisory," Dominic Adams (2014). "Flint Officials Say 'Abnormal' Test to Blame in E. Coli Scare, Water Boil Advisory Remains," *Michigan Live*, August 18, accessed December 11, 2017, http://www.mlive.com/news/flint/index.ssf/2014/08/flint_officials_say_abnormal_t.html.

27. Rick Snyder (2016): 40.

28. Ron Fonger (2015), "Detroit Offers Flint 'Long-Term' Deal for Lake Huron Water with No Reconnection Fee," *Michigan Live*, January 16, accessed December 11, 2017, http://www.mlive.com/news/flint/index.ssf/2015/01/detroit_offers_flint_deal_for.html.

29. Anurag Mantha and Siddhartha Roy (2015), "Hazardous Waste Levels of Lead Found in Flint Household's Water," *Flint Water Study*, 24 August 2015, accessed May 12, 2017, http://flintwaterstudy.org/2015/08/hazardous-waste-levels-of-lead-found-in-a-flint-households-water/; Marc Edwards (2015), "Our Sampling of 252 Homes Demonstrates a High Lead in Water Risk: Flint Should Be Failing to Meet the EPA Lead and Copper Rule," *Flint Water Study*, 8 September, accessed 12 May 2017, http://flintwaterstudy.org/2015/09/our-sampling-of-252-homes-demonstrates-a-high-lead-in-water-risk-flint-should-be-failing-to-meet-the-epa-lead-and-copper-rule/.

30. Rick Snyder (2016): 71.

31. United States Environmental Protection Agency Office of Enforcement and Compliance Assurance, Washington D.C., accessed December 20, 2017, https://www.epa.gov/sites/production/files/2016-01/documents/1_21_sdwa_1431_emergency_admin_order_012116.pdf

32. Josh Sanburn (2017), "Flint's Water Crisis Still Isn't Over Yet. Here's Where Things Stand a Year Later," *Time Magazine*, January 18, accessed December 20, 2017, http://time.com/4634937/flint-water-crisis-criminal-charges-bottled-water/.

33. Michigan Department of Civil Rights (2017), *The Flint Water Crisis: Systemic Racism through the Lens of Flint. Report of the Michigan Civil Rights Commission*, 17 February: 4–5, accessed 12 May 2017, http://www.michigan.gov/documents/mdcr/VFlintCrisisRep-F-Edited3-13-17_554317_7.pdf.

34. Office of the High Commission, "Your Human Rights," *United Nations Human Rights*, accessed December 20, 2017, http://www.ohchr.org/EN/Issues/Pages/WhatareHumanRights.aspx.

35. Ibid.

36. Gary H. Wolff and Meena Palaniappan (2004), "Public or Private Water Management? Cutting the Gordian Knot," *Journal of Water Resources Planning and Management* 130, no. 1 (January/February): 1.

37. Wolff and Palaniappan (2004): 1.

38. Food and Water Watch (2015), "Water Privatization Facts and Figures: Privatizing Local Water and Sewer Systems Usually Does Far More Harm Than Good for Our Communities," 31 August, accessed May 11, 2017, https://www.foodandwaterwatch.org/insight/water-privatization-facts-and-figures.

bibliography

Adams, Dominic (2014). "Flint Officials Say "Abnormal" Test to Blame in E. Coli Scare, Water Boil Advisory Remains," *Michigan Live*, August 18. Accessed December 11, 2017. http://www.mlive.com/news/flint/index.ssf/2014/08/flint_officials_say_abnormal_t.html.

Bay Institute (2016). *San Francisco Bay Freshwater-Starved Estuary: How Water Flowing to the Ocean Sustains California's Greatest Aquatic Ecosystem*. September. Accessed 10 May 2017. https://www.dropbox.com/s/hyjm3r2jp9e09ue/Freshwater_starved_estuary_FullReport.pdf?dl=0.

Bullard, Robert (2000). *Dumping in Dixie: Race, Class and Environmental Quality*, 3rd edition. Boulder: Westview Press.

Campbell, Carla, Rachael Greenberg, Deepa Mankikar, and Ronald Ross (2016). "A Case Study of Environmental Injustice: The Failure in Flint." *International Journal of Environmental Research and Public Health* 13, no. 10 (October): 1–11.

Center on Poverty and Inequality (2011). "20 Facts About US Inequality Everyone Should Know," Stanford University, January 26. http://inequality.stanford.edu/publications/20-facts-about-us-inequality-everyone-should-know.

Chapman, C., J. Laird, N. Ifill, and A. KewalRamani (2011). *Trends in High School Dropout and Completion Rates in the United States: 1972–2009*. (NCES 2012-06). Washington, D.C.: National Center for Education Statistics, Institute of Education Sciences, U.S. Department of Education. Accessed May 9, 2017. http://nces.ed.gov/pubsearch/pubsinfo.asp?pubid=2012006.

Edwards, Marc (2015). "Our Sampling of 252 Homes Demonstrates a High Lead in Water Risk: Flint Should be Failing to Meet the EPA Lead and Copper Rule." *Flint Water Study*, 8 September. Accessed May 12, 2017. http://flintwaterstudy.org/2015/09/our-sampling-of-252-homes-demonstrates-a-high-lead-in-water-risk-flint-should-be-failing-to-meet-the-epa-lead-and-copper-rule/.

Fonger, Ron (2014). "State Says Flint River Water Meets All Standards But More Than Twice the Hardness of Lake Water." *Michigan Live*, May 23. Accessed December 11, 2017. http://www.mlive.com/news/flint/index.ssf/2014/05/state_says_flint_river_water_m.html.

Fonger, Ron (2015). "Detroit Offers Flint 'Long-Term' Deal for Lake Huron Water with No Reconnection Fee." *Michigan Live*, January 16. Accessed December 11, 2017. http://www.mlive.com/news/flint/index.ssf/2015/01/detroit_offers_flint_deal_for.html.

Food and Water Watch (2015). "Water Privatization Facts and Figures: Privatizing Local Water and Sewer Systems Usually Does Far More Harm Than Good for Our Communities." 31 August. Accessed May 11, 2017. https://www.foodandwaterwatch.org/insight/water-privatization-facts-and-figures.

Gregson, Jonathon (2018). "The Richest Countries in the World." *Global Finance*, March 1. Accessed May 12, 2017. https://www.gfmag.com/global-data/economic-data/richest-countries-in-the-world?page=12.

Greenberg, Michael (2016). "Delivering Fresh Water: Critical Infrastructure, Environmental Justice, and Flint, Michigan." *American Journal of Public Health* 106, no. 8 (August): 138–1360.

Hanna-Attisha, Mona, Jenny LaChance, Richard Casey-Sadler, and Allison Champney Schnepp (2016). "Elevated Blood Lead Levels in Children Associated with the Flint Drinking Water Crisis: A Spatial Analysis of Risk and Public Health Response." *American Journal of Public Health* 106, no. 2 (February): 283–90.

Henry, Megan, Rian Watt, Lily Rosenthal, and Azim Shivji, Abt Associates (2016). *The 2016 Annual Homeless Assessment Report to Congress*, The US Department of Housing and Urban Development (November): 1. Accessed May 11, 2017. https://www.hudexchange.info/resources/documents/2016-AHAR-Part-1.pdf.

Julius Wilson, William (2010). "Why Both Social Structure and Culture Matter in a Holistic Analysis of Inner-City Poverty." *Annals of the American Academy of Political and Social Science* 629, no. 1 (May): 200–19.

Kennedy, Merrit (2016). "Lead-Laced Water in Flint: A Step by Step Look at the Makings of a Crisis." *National Public Radio*, April 20. Accessed December 21, 2017. https://www.npr.org/sections/thetwo-way/2016/04/20/465545378/lead-laced-water-in-flint-a-step-by-step-look-at-the-makings-of-a-crisis.

Mantha, Anurag, and Siddhartha Roy (2015). "Hazardous Waste Levels of Lead Found in Flint Household's Water." *Flint Water Study*, 24 August 2015. Accessed May 12, 2017. http://flintwaterstudy.org/2015/08/hazardous-waste-levels-of-lead-found-in-a-flint-households-water/.

Michigan Department of Civil Rights (2017). *The Flint Water Crisis: Systemic Racism through the Lens of Flint. Report of the Michigan Civil Rights Commission*, February 17: 4–5. Accessed May 12, 2017. http://www.michigan.gov/documents/mdcr/VFlintCrisisRep-F-Edited3-13-17_554317_7.pdf.

Office of the High Commission. "Your Human Rights." *United Nations Human Rights.* Accessed December 20, 2017. http://www.ohchr.org/EN/Issues/Pages/WhatareHumanRights.aspx.

Rumberger, Russell (2011). *Dropping Out: Why Students Drop Out of High School and What Can Be Done About It.* Cambridge, MA: Harvard University Press.

Sampson, Robert J., Jeffrey D. Morenoff, and Thomas Gannon-Rowley (2002). "Assessing 'Neighborhood Effects': Social Processes and New Directions in Research." *Annual Review of Sociology* 28 (August): 443–78.

Sanburn, Josh. (2017). "Flint's Water Crisis Still Isn't Over Yet. Here's Where Things Stand a Year Later." *Time Magazine*, January 18. Accessed December 20, 2017. http://time.com/4634937/flint-water-crisis-criminal-charges-bottled-water/.

Sommeiller, Estelle, and Mark Price (2015). "The Unequal States of America: Income Inequality by State 1917–2012." *Economic Policy Institute*, January 26. Accessed May 11, 2017. http://www.epi.org/publication/income-inequality-by-state-1917-to-2012/.

Snyder, Rick (2016). "State of Michigan Executive Office: Lansing," 16. Accessed December 10, 2017. https://www.documentcloud.org/documents/2696071-Snyder-Emails.html#document/p16/a272881.

Turney, Kristin (2010). "Neighborhood Disadvantage, Residential Stability, and Perceptions of Instrumental Support among New Mothers." *Journal of Family Issues* 31, no. 4: 499–524.

United Nations (2017). Progress of Goal 6. Accessed December 21, 2017. https://sustainabledevelopment.un.org/sdg6.

United Nations General Assembly (2010). Resolution 64/292 (3 August). Accessed May 11, 2017. http://www.un.org/es/comun/docs/?symbol=A/RES/64/292&lang=E.
UNICEF and World Health Organization (2015). "25 Years: Progress on Sanitation and Drinking Water: 2015 Update and MDG Assessment," 11. Accessed December 19, 2017. http://files.unicef.org/publications/files/Progress_on_Sanitation_and_Drinking_Water_2015_Update_.pdf.

United States Environmental Protection Agency (2013). *Drinking Water Infrastructure Needs Survey and Assessment: Fifth Report to Congress* (April): 4. Accessed May 10, 2017. https://www.epa.gov/sites/production/files/2015-07/documents/epa816r13006.pdf.

United States Environmental Protection Agency Office of Enforcement and Compliance Assurance, Washington, D.C. Accessed December 20, 2017. https://www.epa.gov/sites/production/files/2016-01/documents/1_21_sdwa_1431_emergency_admin_order_012116.pdf.

Wolff, Gary, and Meena Palaniappan (2004). "Public or Private Water Management? Cutting the Gordian Knot." *Journal of Water Resources Planning and Management* 130, no. 1 (January/February): 1–3.

Bow River falls, Calgary, Alberta. 1920. Glenbow Archives, PO-Calgary-3-25, Calgary, AB.

appendix a:

Resolution adopted by the General Assembly on 28 July 2010
[without reference to a Main Committee (A/64/L.63/Rev.1 and Add.1)]

64/292

UN declaration on the human right to water & sanitation

the general assembly,

recalling its resolutions 54/175 of 17 December 1999 on the right to development, 55/196 of 20 December 2000, by which it proclaimed 2003 the International Year of Freshwater, 58/217 of 23 December 2003, by which it proclaimed the International Decade for Action, "Water for Life", 2005–2015, 59/228 of 22 December 2004, 61/192 of 20 December 2006, by which it proclaimed 2008 the International Year of Sanitation, and 64/198 of 21 December 2009 regarding the midterm comprehensive review of the implementation of the International Decade for Action, "Water for Life"; Agenda 21 of June 1992;[1] the Habitat Agenda of 1996;[2] the Mar del Plata Action Plan of 1977 adopted by the United Nations Water Conference;[3] and the Rio Declaration on Environment and Development of June 1992,[4]

recalling also the Universal Declaration of Human Rights,[5] the International Covenant on Economic, Social and Cultural Rights,[6] the International Covenant on Civil and Political Rights,[6] the International Convention on the Elimination of All Forms of Racial Discrimination,[7] the Convention on the Elimination of All Forms of Discrimination against Women,[8] the Convention on the Rights of the Child,[9] the Convention on the Rights of Persons with Disabilities[10] and the Geneva Convention relative to the Protection of Civilian Persons in Time of War, of 12 August 1949,[11]

Recalling further all previous resolutions of the Human Rights Council on human rights and access to safe drinking water and sanitation, including Council resolutions 7/22 of 28 March 2008[12] and 12/8 of 1 October 2009,[13] related to the human right to safe and clean drinking water and sanitation, general comment No. 15 (2002) of the Committee on Economic, Social and Cultural Rights, on the right to water (articles 11 and 12 of the International Covenant on Economic, Social[14] and Cultural Rights)13F and the report of the United Nations High Commissioner for Human Rights on the scope and content of the relevant human rights obligations related to equitable access to safe drinking water and sanitation under international[15] human rights instruments, 14F as well as the report of the independent expert on the issue of human rights obligations related to access to safe drinking water and sanitation,[16]

deeply concerned that approximately 884 million people lack access to safe drinking water and that more than 2.6 billion do not have access to basic sanitation, and alarmed that approximately 1.5 million children under 5 years of age die and 443 million school days are lost each year as a result of water and sanitation-related diseases,

acknowledging the importance of equitable access to safe and clean drinking water and sanitation as an integral component of the realization of all human rights,

reaffirming the responsibility of States for the promotion and protection of all human rights, which are universal, indivisible, interdependent and interrelated, and must be treated globally, in a fair and equal manner, on the same footing and with the same emphasis,

Bearing in mind the commitment made by the international community to fully achieve the Millennium Development Goals, and stressing, in that context, the resolve of Heads of State and Government, as expressed in the United Nations[17] Millennium Declaration, 16F to halve, by 2015, the proportion of people who are unable to reach or afford safe drinking water and, as agreed in the Plan of Implementation of the World Summit on Sustainable Development ("Johannesburg[18] Plan of Implementation"),17F to halve the proportion of people without access to basic sanitation,

1. recognizes the right to safe and clean drinking water and sanitation as a human right that is essential for the full enjoyment of life and all human rights;

2. calls upon States and international organizations to provide financial resources, capacity-building and technology transfer, through international assistance and cooperation, in particular to developing countries, in order to scale up efforts to provide safe, clean, accessible and affordable drinking water and sanitation for all;

3. welcomes the decision by the Human Rights Council to request that the independent expert on human rights obligations related to access to safe drinking water and sanitation submit an annual report to the General Assembly, 13 and encourages her to continue working on all aspects of her mandate and, in consultation with all relevant United Nations agencies, funds and programmes, to include in her report to the Assembly, at its sixty-sixth session, the principal challenges related to the realization of the human right to safe and clean drinking water and sanitation and their impact on the achievement of the Millennium Development Goals.

108th plenary meeting, 28 July 2010

notes

1. *Report of the United Nations Conference on Environment and Development, Rio de Janeiro, 3-14 June 1992*, vol. I, *Resolutions Adopted by the Conference* (United Nations publication, Sales No. E.93.I.8 and corrigendum), resolution 1, annex II
2. *Report of the United Nations Conference on Human Settlements (Habitat II), Istanbul, 3-14 June 1996* (United Nations publication, Sales No. E.97.IV.6), chap. I, resolution 1, annex II.
3. *Report of the United Nations Water Conference, Mar del Plata, 14–25 March 1977* (United Nations publication, Sales No. E.77.II.A.12), chap. I.
4. *Report of the United Nations Conference on Environment and Development, Rio de Janeiro, 3-14 June 1992*, vol. I, *Resolutions Adopted by the Conference* (United Nations publication, Sales No. E.93.I.8 and corrigendum), resolution 1, annex I.
5. Resolution 217 A (III).
6. See resolution 2200 A (XXI), annex.
7. United Nations, *Treaty Series*, vol. 660, No. 9464.
8. Ibid., vol. 1249, No. 20378.
9. Ibid., vol. 1577, No. 27531.
10. Resolution 61/106, annex I.
11. United Nations, *Treaty Series*, vol. 75, No. 973.
12. See *Official Records of the General Assembly, Sixty-third Session, Supplement No. 53* (A/63/53), chap. II.
13. See A/HRC/12/50 and Corr.1, part one, chap. I.
14. See *Official Records of the Economic and Social Council, 2003, Supplement No. 2* (E/2003/22), annex IV.
15. A/HRC/6/3.
16. A/HRC/12/24.
17. See resolution 55/2.
18. *See Report of the World Summit on Sustainable Development, Johannesburg, South Africa, 26 August–4 September 2002* (United Nations publication, Sales No. E.03.II.A.1 and corrigendum), chap. I, resolution 2, annex.

appendix b:

excerpts from: **UN declaration on the rights of Indigenous peoples**

[107 Plenary Meeting, 13 December 2007]

the general assembly,

guided by the purposes and principles of the Charter of the United Nations, and good faith in the fulfillment of the obligations assumed by States in accordance with the Charter,

affirming that indigenous peoples are equal to all other peoples, while recognizing the right of all peoples to be different, to consider themselves different, and to be respected as such,

affirming also that all peoples contribute to the diversity and richness of civilizations and cultures, which constitute the common heritage of humankind,

. . .

concerned that indigenous peoples have suffered from historic injustices as a result of, inter alia, their colonization and dispossession of their lands, territories and resources, thus preventing them from exercising, in particular, their right to development in accordance with their own needs and interests, *recognizing* the urgent need to respect and promote the inherent rights of indigenous peoples which derive from their political, economic and social structures and from their cultures, spiritual traditions, histories and philosophies, especially their rights to their lands, territories and resources,

. . .

welcoming the fact that indigenous peoples are organizing themselves for political, economic, social and cultural enhancement and in order to bring to an end all forms of discrimination and oppression wherever they occur,

convinced that control by indigenous peoples over developments affecting them and their lands, territories and resources will enable them to maintain and strengthen their institutions, cultures and traditions, and to promote their development in accordance with their aspirations and needs,

recognizing that respect for indigenous knowledge, cultures and traditional practices contributes to sustainable and equitable development and proper management of the environment,

. . .

recognizing and reaffirming that indigenous individuals are entitled without discrimination to all human rights recognized in international law, and that indigenous peoples possess collective rights which are indispensable for their existence, well-being and integral development as peoples,

solemnly proclaims the following United Nations Declaration on the Rights of Indigenous Peoples as a standard of achievement to be pursued in a spirit of partnership and mutual respect:

article 1

Indigenous peoples have the right to the full enjoyment, as a collective or as individuals, of all human rights and fundamental freedoms as recognized in the Charter of the United Nations, the Universal Declaration of Human Rights and international human rights law.

article 2

Indigenous peoples and individuals are free and equal to all other peoples and individuals and have the right to be free from any kind of discrimination, in the exercise of their rights, in particular that based on their indigenous origin or identity.

article 3

Indigenous peoples have the right to self-determination. By virtue of that right they freely determine their political status and freely pursue their economic, social and cultural development.

article 4

Indigenous peoples, in exercising their right to self-determination, have the right to autonomy or self-government in matters relating to their internal and local affairs, as well as ways and means for financing their autonomous functions.

article 5

Indigenous peoples have the right to maintain and strengthen their distinct political, legal, economic, social and cultural institutions, while retaining their right to participate fully, if they so choose, in the political, economic, social and cultural life of the State.

article 7

1. Indigenous individuals have the rights to life, physical and mental integrity, liberty and security of person.

2. Indigenous peoples have the collective right to live in freedom, peace and security as distinct peoples and shall not be subjected to any act of genocide or any other act of violence, including forcibly removing children of the group to another group.

article 9

Indigenous peoples and individuals have the right to belong to an indigenous community or nation, in accordance with the traditions and customs of the community or nation concerned. No discrimination of any kind may arise from the exercise of such a right.

article 10

Indigenous peoples shall not be forcibly removed from their lands or territories. No relocation shall take place without the free, prior and informed consent of the indigenous peoples concerned and after agreement on just and fair compensation and, where possible, with the option of return.

article 11

1. Indigenous peoples have the right to practise and revitalize their cultural traditions and customs. This includes the right to maintain, protect and develop the past, present and future manifestations of their cultures, such as archaeological and historical sites, artefacts, designs, ceremonies, technologies and visual and performing arts and literature.

2. States shall provide redress through effective mechanisms, which may include restitution, developed in conjunction with indigenous peoples, with respect to their cultural, intellectual, religious and spiritual property taken without their free, prior and informed consent or in violation of their laws, traditions and customs.

article 12

1. Indigenous peoples have the right to manifest, practise, develop and teach their spiritual and religious traditions, customs and ceremonies; the right to maintain, protect, and have access in privacy to their religious and cultural sites; the right to the use and control of their ceremonial objects; and the right to the repatriation of their human remains.

2. States shall seek to enable the access and/or repatriation of ceremonial objects and human remains in their possession through fair, transparent and effective mechanisms developed in conjunction with indigenous peoples concerned.

article 13

1. Indigenous peoples have the right to revitalize, use, develop and transmit to future generations their histories, languages, oral traditions, philosophies, writing systems and literatures, and to designate and retain their own names for communities, places and persons.

2. States shall take effective measures to ensure that this right is protected and also to ensure that indigenous peoples can understand and be understood in political, legal and administrative proceedings, where necessary through the provision of interpretation or by other appropriate means.

article 20

1. Indigenous peoples have the right to maintain and develop their political, economic and social systems or institutions, to be secure in the enjoyment of their own means of subsistence and development, and to engage freely in all their traditional and other economic activities.

2. Indigenous peoples deprived of their means of subsistence and development are entitled to just and fair redress.

article 21

1. Indigenous peoples have the right, without discrimination, to the improvement of their economic and social conditions, including, inter alia, in the areas of education, employment, vocational training and retraining, housing, sanitation, health and social security.

2. States shall take effective measures and, where appropriate, special measures to ensure continuing improvement of their economic and social conditions. Particular attention shall be paid to the rights and special needs of indigenous elders, women, youth, children and persons with disabilities.

article 23

Indigenous peoples have the right to determine and develop priorities and strategies for exercising their right to development. In particular, indigenous peoples have the right to be actively involved in developing and determining health, housing and other economic and social programmes affecting them and, as far as possible, to administer such programmes through their own institutions.

article 24

1. Indigenous peoples have the right to their traditional medicines and to maintain their health practices, including the conservation of their vital medicinal plants, animals and minerals. Indigenous individuals also have the right to access, without any discrimination, to all social and health services.

2. Indigenous individuals have an equal right to the enjoyment of the highest attainable standard of physical and mental health. States shall take the necessary steps with a view to achieving progressively the full realization of this right.

article 25

Indigenous peoples have the right to maintain and strengthen their distinctive spiritual relationship with their traditionally owned or otherwise occupied and used lands, territories, waters and coastal seas and other resources and to uphold their responsibilities to future generations in this regard.

article 26

1. Indigenous peoples have the right to the lands, territories and resources which they have traditionally owned, occupied or otherwise used or acquired.

2. Indigenous peoples have the right to own, use, develop and control the lands, territories and resources that they possess by reason of traditional ownership or other traditional occupation or use, as well as those which they have otherwise acquired.

3. States shall give legal recognition and protection to these lands, territories and resources. Such recognition shall be conducted with due respect to the customs, traditions and land tenure systems of the indigenous peoples concerned.

article 29

1. Indigenous peoples have the right to the conservation and protection of the environment and the productive capacity of their lands or territories and resources. States shall establish and implement assistance programmes for indigenous peoples for such conservation and protection, without discrimination.

2. States shall take effective measures to ensure that no storage or disposal of hazardous materials shall take place in the lands or territories of indigenous peoples without their free, prior and informed consent.

3. States shall also take effective measures to ensure, as needed, that programmes for monitoring, maintaining and restoring the health of indigenous peoples, as developed and implemented by the peoples affected by such materials, are duly implemented.

article 31

1. Indigenous peoples have the right to maintain, control, protect and develop their cultural heritage, traditional knowledge and traditional cultural expressions, as well as the manifestations of their sciences, technologies and cultures, including human and genetic resources, seeds, medicines, knowledge of the properties of fauna and flora, oral traditions, literatures, designs, sports and traditional games and visual and performing arts. They also have the right to maintain, control, protect and develop their intellectual property over such cultural heritage, traditional knowledge, and traditional cultural expressions.

article 32

1. Indigenous peoples have the right to determine and develop priorities and strategies for the development or use of their lands or territories and other resources.

2. States shall consult and cooperate in good faith with the indigenous peoples concerned through their own representative institutions in order to obtain their free and informed consent prior to the approval of any project affecting their lands or territories and other resources, particularly in connection with the development, utilization or exploitation of mineral, water or other resources.

3. States shall provide effective mechanisms for just and fair redress for any such activities, and appropriate measures shall be taken to mitigate adverse environmental, economic, social, cultural or spiritual impact.

article 37

1. Indigenous peoples have the right to the recognition, observance and enforcement of treaties, agreements and other constructive arrangements concluded with States or their successors and to have States honour and respect such treaties, agreements and other constructive arrangements.

2. Nothing in this Declaration may be interpreted as diminishing or eliminating the rights of indigenous peoples contained in treaties, agreements and other constructive arrangements.

article 38

States, in consultation and cooperation with indigenous peoples, shall take the appropriate measures, including legislative measures, to achieve the ends of this Declaration.

article 39

Indigenous peoples have the right to have access to financial and technical assistance from States and through international cooperation, for the enjoyment of the rights contained in this Declaration.

article 40

Indigenous peoples have the right to access to and prompt decision through just and fair procedures for the resolution of conflicts and disputes with States or other parties, as well as to effective remedies for all infringements of their individual and collective rights. Such a decision shall give due consideration to the customs, traditions, rules and legal systems of the indigenous peoples concerned and international human rights.

article 43

The rights recognized herein constitute the minimum standards for the survival, dignity and well-being of the indigenous peoples of the world.

article 44

All the rights and freedoms recognized herein are equally guaranteed to male and female indigenous individuals.

article 45

Nothing in this Declaration may be construed as diminishing or extinguishing the rights indigenous peoples have now or may acquire in the future.

article 46

1. Nothing in this Declaration may be interpreted as implying for any State, people, group or person any right to engage in any activity or to perform any act contrary to the Charter of the United Nations or construed as authorizing or encouraging any action which would dismember or impair, totally or in part, the territorial integrity or political unity of sovereign and independent States.

2. In the exercise of the rights enunciated in the present Declaration, human rights and fundamental freedoms of all shall be respected. The exercise of the rights set forth in this Declaration shall be subject only to such limitations as are determined by law and in accordance with international human rights obligations. Any such limitations shall be non-discriminatory and strictly necessary solely for the purpose of securing due recognition and respect for the rights and freedoms of others and for meeting the just and most compelling requirements of a democratic society.

3. The provisions set forth in this Declaration shall be interpreted in accordance with the principles of justice, democracy, respect for human rights, equality, non-discrimination, good governance and good faith.

Glenmore Dam, Calgary, Alberta. 1935. Glenbow Archives, PA-4076-81.

contributors

Warren Cariou was born into a family of Métis and European ancestry in Meadow Lake, Saskatchewan. He has published works of fiction and memoir as well as critical writing about Indigenous storytelling, literature, and environmental philosophy. Cariou has also created two films about Indigenous communities in Western Canada's tar sands region, and has written numerous articles, stories, and poems about Indigeneity and petroleum. His visual art project, *Petrography*, uses tar sands bitumen as a photographic medium. Cariou is a professor of English at the University of Manitoba, where he directs the Centre for Creative Writing and Oral Culture.

Michelle Daigle is Mushkegowuk (Swampy Cree) and of French ancestry. She is a member of Constance Lake First Nation located in Treaty 9 territory, and an assistant professor in the Department of Geography at the University of British Columbia. She is interested in bringing the study of geography into critical dialogue with Indigenous Studies to examine colonial-capitalist dispossession — specifically exploitative extractive developments — as well as Indigenous movements for decolonization and self-determination.

Jim Ellis is a professor of English and director of the Calgary Institute for the Humanities. He is the author of *Sexuality and Citizenship* (2003) and *Derek Jarman's Angelic Conversations* (2009), and the editor of the previous volume in the CIH's Community Seminar series, *Calgary, City of Animals* (University of Calgary Press, 2017).

Flora Giesbrecht is the coordinator of the Elbow River Watershed Partnership.

Dr. Jodi Hilty is an expert on wildlife corridors, and is the president and chief scientist of the Yellowstone to Yukon Conservation Initiative. For over twenty years she has worked to advance conservation by leading science and community-based and collaborative conservation to advance policy and management. In the last fifteen years her work has focused on North America. Hilty has been co-editor or lead author on three books, most recently *Climate and Conservation: Landscape and Seascape Science, Planning, and Action*. She currently serves on the Board of the Smith Fellowship and as deputy chair of the IUCN Connectivity Committee.

Helen Knott is an activist and poet-writer working for Indigenous land rights in Canada. She has participated in the Treaty 8 Caravan across Canada — a cross-country caravan for justice and peace that travelled to major cities to talk about stopping the Site C Dam in British Columbia — and writes about the relationship between resource extraction and violence against Indigenous women. Knott is Dane Zaa and Nehiyawak from the Prophet River First Nation in British Columbia. In 2017, she was selected to be part of the Nobel Women's Initiative.

David K. Laidlaw is a Research Fellow with the Canadian Institute of Resources Law, with a B.Sc. in Computer Science and Economics from the University of Calgary in 1985 and an LL.B. from Dalhousie University in 1988. Admitted to the Law Society of Alberta in 1989, after twenty years of private practice in Calgary he returned to law school and obtained an LL.M. from the University of Calgary in 2013. Having practiced Aboriginal law in the early 1990s, Laidlaw's ongoing interest in the area has led to research focusing on Aboriginal law. Other interests include environmental law, legal history, and jurisprudence.

Ciara McKeown is a public art commissioner, curator, director, producer, and project manager who has worked with multiple arts organizations across Canada and the US for over a decade. She has recently started her own practice writing, researching, and working with clients on public art planning. Ciara is an executive board member with Public Art Dialogue, and was co-organizer of "Public Art: New Ways of Thinking and Working," a symposium hosted by York University in May 2017. The first of its kind in Canada, the symposium invited cross-disciplinary perspectives and research to critically examine the current state of Canadian contemporary public art practices. Ciara holds an M.A. from New York University and a B.A. from McGill University.

Josée Méthot is an environmental scientist and process designer with a passion for working on complex problems in the water-energy-food nexus. She is the planning manager with the Red Deer River Watershed Alliance where she collaborates with community and research partners to advance a mix of practical and ambitious solutions to water issues. Méthot holds an MSc. in Natural Resource Sciences from McGill University and currently teaches topics related to systems change and social innovation at Mount Royal University. She loves to explore beautiful landscapes throughout Western Canada and is working on her paddling skills.

Adrian Parr is an internationally recognized environmental, political, and cultural thinker and activist. She is the director of the University of Cincinnati Taft Research Center and a UNESCO Co-Chair of Water Access and Sustainability. She is a founding signatory of the *Geneva Actions on Water Security*. In 2011 Parr received the UC Rieveschl Award for Scholarly and Creative Work.

Amy Spark is an environmental scientist and advocate specializing in the intersection between ecological and mental health. She holds an MSc. with Distinction in Environment, Culture & Society from the University of Edinburgh and currently works at Alberta Ecotrust Foundation and at Bow Valley College. Spark is the co-founder of Refugia Retreats, an initiative that acts as a catalyst for societal and personal change through retreats, workshops, and facilitation. She is an amateur urban homesteader, and an outdoors enthusiast, budding writer, and lover of science fiction.

Leslie Sweder is a practicing artist living in Calgary. She has developed a multidisciplinary practice which includes public interventions, painting, collaborative drawing and photography. Each of these practices are direct responses to and/or recordings of her surrounding environment and each one implicitly feeds the others. Her public interventions have been supported by both the AFA and Calgary 2012, and are featured in the upcoming documentary, *Sidewalk Citizen*. Sweder is one-third of the collaborative trio *Drunken Paw*, whose work is in numerous collections in North America including the Alberta Foundation for the Arts.

Charles Tepperman is an associate professor in the Department of Communication, Media and Film at the University of Calgary. He has published articles on Canadian film policy, non-theatrical film culture, film technology, and early cinema in Canada. Tepperman is the author of *Amateur Cinema: The Rise of North American Moviemaking, 1923–1960* (University of California, 2015) and director of the Amateur Movie Database Project.

Nancy Tousley, recipient of a Governor General's Award for Visual and Media Arts for outstanding contribution in 2011, is a senior art critic, arts journalist, and independent curator. Her work has been recognized with many prestigious awards including the Board of Governors of the Alberta College of Art and Design's Award of Excellence (1997), the Ontario Association of Art Galleries award for best curatorial writing on contemporary art (1999 and 2001), the Canadian Museums Association's award for outstanding achievement in arts journalism (2002), and the medal of the Royal Canadian Academy of Art (2009).

a PROUD PARTNER in

Campus Alberta

A book in the **Campus Alberta Collection**, a collaboration of Athabasca University Press, University of Alberta Press, University of Calgary Press

University of Calgary Press
press.ucalgary.ca

Water Rites: Reimagining Water in the West
Edited by Jim Ellis
978-1-55238-997-3 (pb)

Writing Alberta: Building on a Literary Identity
Edited by George Melnyk and Donna Coates
978-1-55238-890-7 (pb)

The Frontier of Patriotism: Alberta and the First World War
Edited by Adriana A. Davies and Jeff Keshen
978-1-55238-834-1 (pb)

So Far and Yet So Close: Frontier Cattle Ranching in Western Prairie Canada and the Northern Territory of Australia
Warren M. Elofson
978-1-55238-794-8 (pb)

Athabasca University Press
aupress.ca

The Medium Is the Monster: Canadian Adaptations of* Frankenstein *and the Discourse of Technology
Mark A. McCutcheon
978-1-77199-236-7 (cl)
978-1-77199-224-4 (pb)

Public Deliberation on Climate Change: Lessons from Alberta Climate Dialogue
Edited by Lorelei L. Hanson
978-1-77199-215-2 (pb)

Visiting With the Ancestors: Blackfoot Shirts in Museum Spaces
Laura Peers and Alison K. Brown
978-1-77199-037-0 (pb)

Alberta Oil and the Decline of Democracy in Canada
Edited by Meenal Shrivastava and Lorna Stefanick
978-1-77199 029-5 (pb)

University of Alberta Press
uap.ualberta.ca

Keetsahnak / Our Missing and Murdered Indigenous Sisters
Kim Anderson, Maria Campbell and Christi Belcourt, Editors
978-1-77212-367-8 (pb)

Trudeau's Tango: Alberta Meets Pierre Elliott Trudeau, 1968–1972
Darryl Raymaker
978-1-77212-265-7 (pb)

Seeking Order in Anarchy: Multilateralism as State Strategy
Robert W. Murray, Editor
978 1-77212-139-1 (pb)

Upgrading Oilsands Bitumen and Heavy Oil
Murray R. Gray
978 1 77212-035-6 (hc)